CHEMICAL LECTURE EXPERIMENTS

CHEMICAL LECTURE

EXPERIMENTS

NON-METALLIC ELEMENTS

BY

G. S. NEWTH, F.I.C., F.C.S.

FORMERLY CHEMICAL LECTURE DEMONSTRATOR IN THE ROYAL COLLEGE OF SCIENCE
SOUTH KENSINGTON

NEW IMPRESSION

LONGMANS, GREEN, AND CO.
39 PATERNOSTER ROW, LONDON
55 FIFTH AVENUE, NEW YORK
BOMBAY, CALCUTTA, AND MADRAS

1922

PREFACE

The object of this book is twofold. Firstly, it is intended to supply chemical lecturers and teachers with a useful repertoire of experiments, suitable for illustrating upon the lecture table the modes of preparation, and the properties, of the non-metallic elements and their commoner and more important compounds. I have therefore given such full directions for the preparation and performance of the various experiments described, as will enable any experimenter to successfully repeat them. No mere description of experiments, however minute, can entirely take the place of experience, but it has been my endeavour throughout this book to give such details, resulting from my own experience, as shall meet as far as possible the lack of experience in others.

No account of any experiment has been introduced into the book upon the authority solely of any verbal or printed description, but every experiment has been the subject of my personal investigation, and illustrated in every case, with the exception of three,[1] by woodcuts made from original drawings.

What should be and what should not be included under the head of 'experiments suitable for illustration upon the lecture table' may be a matter for difference of opinion, and must depend often upon circumstances. I have excluded for the most part all such experiments as are merely illustrative of the ordinary laboratory processes of qualitative analysis; also all experiments illustrating the action of substances upon animal

[1] Figs. 1 and 24, 'Inorganic Chemistry,' Frankland and Japp; fig. 203, 'Ganot's Physics.'

life, and, lastly, such as are of either so critical or dangerous a nature as to involve the adoption of any special or elaborate precautions.

Throughout the book the French metrical system has been adopted, but as the English mind, even of scientific men, often continues to conceive of measures of length in inches, long after it has acquired the habit of thinking in grams and cubic centimetres, I have added, when deemed desirable, the English equivalent of the French measures of length. It will be convenient to remember that for all practical purposes 25 millimetres, or 2·5 centimetres, are equal to one inch.

For the convenience of such teachers as may not have ready access to books of reference, I have added, in the form of an appendix, a number of important and useful tables.

Secondly, it is my object to furnish the chemical student with a book which shall serve as a companion to the lectures he may attend—a book in which he will find fully described most, if not all, of the experiments he is likely to see performed upon the lecture table, and which will therefore relieve him from the necessity of laboriously noting them and often sketching the apparatus used. In this way the student will be spared much unnecessary and distracting work during the lecture, and will therefore be better able to give his undivided attention to the explanations or arguments of the lecturer.

Further, to meet the wants of the chemical student, I have given the equations representing the various reactions which are described in the book; and, although this work is not designed to take the place of any existing text-book, it has been so arranged that the student may learn from it the methods of preparation and most of the important properties of the non-metallic elements and their more common compounds.

G. S. N.

ROYAL COLLEGE OF SCIENCE, SOUTH KENSINGTON, W :
September 1892.

CONTENTS

—◆—

	PAGE
HYDROGEN	1
HYDROGENIUM . . .	18
OXYGEN	20
OZONE	29
WATER	38
HYDROGEN PEROXIDE . .	74
CHLORINE	79
HYDROCHLORIC ACID . .	94
OXIDES AND ACIDS OF CHLORINE	100
BROMINE	105
HYDROBROMIC ACID . .	106
IODINE	110
HYDRIODIC ACID . .	113
OXIDES AND OXYACIDS OF IODINE	113
FLUORINE	114
HYDROFLUORIC ACID .	115
SILICON FLUORIDE . .	116
NITROGEN	117
ARGON	126
AMMONIA	128
HYDRAZOIC ACID . .	138
NITROUS CHLORIDE . .	141
NITROGEN IODIDE . .	142
OXIDES OF NITROGEN .	143
THE ATMOSPHERE . .	154
DIFFUSION	161
COMBUSTION	167
CARBON	179
CARBON DIOXIDE . .	186
CARBON MONOXIDE . .	197
METHANE (MARSH GAS) .	204

	PAGE
ETHYLENE	207
ACETYLENE	212
FLAME, AND LUMINOSITY OF FLAME	217
SILICON	238
SILICON HYDRIDE . .	238
SILICIC ACID . . .	239
BORON	240
BORIC ACID . . .	241
PHOSPHORUS . . .	242
PHOSPHORETTED HYDROGEN	249
PHOSPHORUS PENTOXIDE .	254
PHOSPHOROUS OXIDE .	255
SULPHUR	258
SULPHURETTED HYDROGEN .	262
HYDROGEN PERSULPHIDE	267
SULPHUR DIOXIDE . .	267
SULPHUR TRIOXIDE . .	271
SULPHURIC ACID . .	273
CARBON DISULPHIDE .	275
ARSENIC	281
ARSENIURETTED HYDROGEN	282
ANTIMONY	283
ANTIMONIURETTED HYDROGEN	285
DISSOCIATION . . .	286
LIQUEFACTION OF GASES .	293
EXPERIMENTS ON ELECTROLYSIS	311
LANTERN ILLUSTRATIONS .	316
APPENDIX . . .	329
INDEX	345

CHEMICAL LECTURE EXPERIMENTS

HYDROGEN

MODES OF PREPARATION

1. By the action of sodium or potassium upon water $Na + H_2O = NaHO + H$. A fragment of sodium or potassium, about the size of a pea, is thrown upon cold water contained in a soup plate. The plate should be instantly covered by a glass bell-jar when potassium is employed, as the molten globule of potassium hydroxide which remains floating, in a spheroidal state, upon the water for a few seconds, will, as it cools, be scattered with some violence out of the plate.

If a fragment of sodium, instead of being thrown upon water, be placed upon a piece of wet filter paper in a plate, the heat of the reaction will rise sufficiently high to inflame the hydrogen, and in this case it must also be covered with a bell-jar.

2. To collect hydrogen from the action of sodium upon water, the following is the best method, and the only one which is free from danger. A piece of ordinary lead or composition pipe, about 25 millimetres long and about 4 millimetres bore, has one end closed by squeezing in a vice, or by a few taps with a hammer. A pellet of sodium is rolled between the fingers and pushed into this tube. It may be well forced in by pressing the mouth of the tube down upon the table, the excess of the metal being afterwards trimmed off with a knife. This tube with its

B

contents is then dropped into the pneumatic trough, in which it
will sink, and hydrogen will be evolved in a steady and gentle
stream. The mouth of an inverted glass cylinder can be brought

FIG. 1.

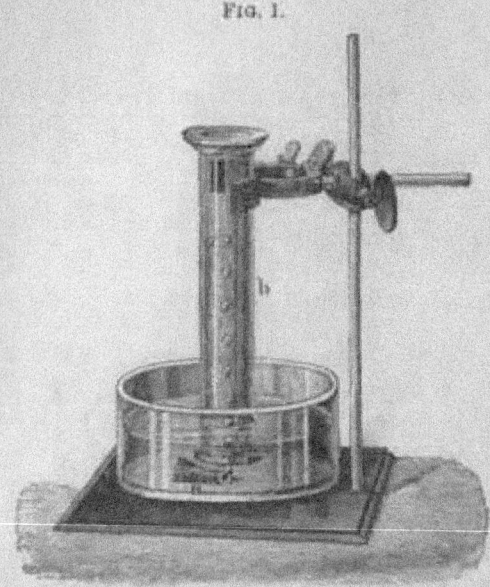

over the stream of
uprising bubbles, and
the gas so collected.
If the little tubes are
filled as described, it
is impossible for any
explosion to take
place in the perform-
ance of this experi-
ment. The filled tubes
may be preserved in
naphtha.

3. By the action
of zinc-copper couple
on water. $Zn + H_2O =
ZnO + H_2$. About 30
grams of zinc-dust is
placed in a small flask
of about 150 c.c. capa-
city, and the flask filled up with a solution of copper sulphate
(containing 40 grams of the salt to the litre). The contents of
the flask are then briskly shaken together, the coppered zinc is
allowed to settle and the water poured off. The flask is then
filled to the neck with distilled water, and a cork carrying a
delivery tube is inserted; on heating the contents of the flask a
slow stream of hydrogen is evolved. Zinc filings may be substituted for zinc-dust; but in this case the action will be slower.

4. By the action of magnesium on steam. $Mg + H_2O
= MgO + H_2$. A strip of magnesium ribbon about 12 centimetres
long is folded up and placed in a bulb, which is blown in the
middle of a piece of combustion tube about 24 centimetres long.
Each end of the tube is fitted with a cork carrying a short piece

of narrow tube. A current of steam is passed through the tube, which should be held in a clamp and inclined slightly downwards towards the incoming steam (fig. 2), and it is well to pass

the steam through a small empty flask immediately before it enters the bulb-tube. While the steam is passing, the bulb-tube should be warmed throughout its whole length by means of a Bunsen flame, until it is hot enough to prevent condensation of the steam, and then

Fig. 2.

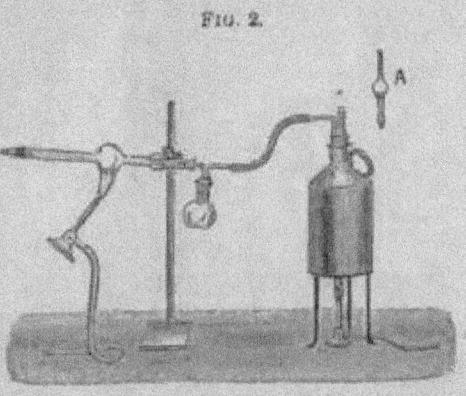

the lamp directed entirely upon the fragment of magnesium within the bulb. As the metal approaches to a red heat it will burst into flame, and if the supply of steam be regulated the hydrogen can be ignited as it issues from the extremity of the tube. The supply of steam may be easily regulated by partially withdrawing the lamp from the vessel in which the steam is being generated.

Instead of fitting corks to the bulb-tube, a most convenient plan is to use pieces of glass tube, so drawn out as to taper at

Fig. 3.

each end. A piece of caoutchouc slipped over such a tube will enable it to fit into tubes of various bores.

These 'connectors' are so useful for a variety of purposes that it will be found convenient to have a number of them, varying in size, ready for use.

A simple form of boiler for supplying steam for lecture

experiments may be readily made in the following way. An ordinary half-gallon tin can is fitted with a cork carrying two glass tubes; one bent at right angles to convey the steam from the boiler; the other and wider tube being straight, and projecting a short distance above the cork. This wider tube is to carry the 'safety valve,' which consists of a bulb-tube of glass, loaded with mercury, until its weight is about 30 to 35 grams (A, Fig. 2). This is ground to fit the tube in which it sits, by applying a little fine emery moistened with turpentine and revolving the bulb-tube for a few turns in its place.

The fact that steam is capable of supporting the combustion of magnesium may be demonstrated by lowering into a large flask, in which a small quantity of water is being rapidly boiled, a piece of burning magnesium, attached to a wire. The magnesium continues to burn, but the fact that hydrogen is evolved cannot be simultaneously shown.

5. Magnesium is capable of decomposing water readily at the ordinary temperature, if the metal be previously dissolved in mercury. Magnesium amalgam is easily prepared by heating in a test tube a quantity of mercury along with fragments of freshly-scraped magnesium ribbon, when, at a temperature somewhat below the boiling-point of mercury, the two metals will unite with moderate energy.

The amalgam may then be placed in a small flask, fitted with a cork carrying a thistle funnel and delivery tube. On introducing water, a rapid evolution of hydrogen takes place, and the gas may be collected at the pneumatic trough.

6. By the action of red-hot iron upon steam. $3Fe + 4H_2O = Fe_3O_4 + 4H_2$. Steam from a boiler is passed through an iron pipe filled with iron nails, heated to bright redness in either a coke furnace or an Erlenmeyer's gas combustion furnace.

7. This experiment may be performed on a smaller scale by employing a platinum tube about 16 or 18 centimetres long, filled with small 'brads,' and heated by two Bunsens side by side, or

better by a flat-flame Bunsen. The ends of the tube may be stopped with glass connectors, which can be fitted in by means of asbestos paper. A thin flake can be peeled from a strip of asbestos millboard and wound round the connector, in lieu of the caoutchouc, and in this way a tight joint will be made which will not suffer from the heat. A platinum tube is easily made by rolling a piece of foil round an iron rod of suitable size, e.g. the stem of a retort stand, and fastening it temporarily with two or three pieces of wire. The overlapping edges of the foil may be welded together by projecting a blowpipe flame upon them, and gently tapping the metal with a hammer. By carefully heating and hammering all along the seam a perfectly gas-tight tube can be made in a few minutes. The ends may be strengthened by welding a double thickness of foil round them.

It is possible to show the decomposition of steam by means of red-hot iron by substituting a glass tube for the platinum, the glass tube being encased in copper gauze, but the uncertainty of the experiment owing to the risk of the tube becoming cracked makes it unsuitable for lecture illustration.

8. By the action of sulphuric acid upon zinc. $Zn + H_2SO_4 = ZnSO_4 + H_2$. A quantity of granulated zinc is placed in a

FIG. 5

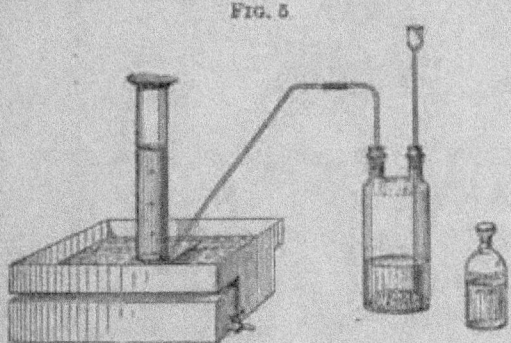

two-necked Woulf's bottle, and covered with water. The Woulf's bottle is fitted with two corks, one carrying a thistle tube which reaches nearly to the bottom of the bottle, the other carrying a tube bent at right angles, which in its turn is connected by means of a piece of caoutchouc tube to a delivery tube: on pouring a little strong acid in through the thistle tube the gas is disengaged.

The evolution of hydrogen will be rendered more prompt and more rapid by adding a few drops of a solution of copper sulphate to the contents of the bottle before introducing the sulphuric acid.

Hydrochloric acid may be substituted for sulphuric acid, and in either case iron can be employed instead of zinc. The use of iron, however, results in a more impure sample of hydrogen, and the method is not so well adapted for lecture purposes.

9. By the action of sodium hydroxide upon zinc. $Zn + 2NaHO = ZnNa_2O_2 + H_2$. Zinc filings are boiled with a strong solution of caustic soda in a small flask fitted with a delivery tube: a slow stream of hydrogen will be evolved. Aluminium filings or turnings may be substituted for zinc.

The best form of pneumatic trough for the lecture table is

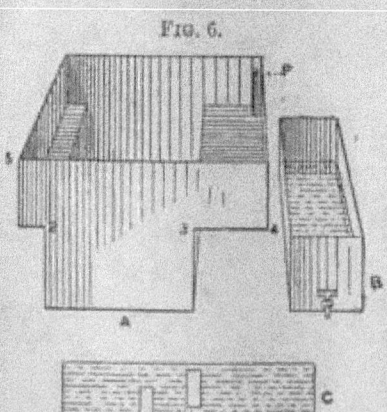

FIG. 6.

that shown in figure 6. It consists of three parts—A, the trough proper, B an overflow vessel, and C the bridge or shelf. The trough is provided with an overflow pipe P, which conveys any excess of water into the vessel B, placed underneath the overhanging portion of the trough; this vessel is also furnished with a stopcock, so that its contents may be led away to a sink. All the parts should be made of stout tinned iron and japanned; for the sake of appearance the inside should be white. In practice two sizes will be found convenient. For the larger, length = 14 in. (35·5 c.m.); width = 9½ in. (24 c.m.); depth = 6 in. (15 c.m.). From 5 to 1 = 2½ in. (6·5 c.m.); from 1 to 2 = 1 in. (2·5 c.m.); from 2 to 3 = 5½ in. (14 c.m.); from 3 to 4 = 3 in. (8 c.m.). Bridge 4½ in. (11·5 c.m.) wide. For the smaller size, length = 10 in. (25·5 c.m.); width = 6 in. (15 c.m.); depth = 3½ in. (9 c.m.). From 1 to 2 = ½ in. (1·25 c.m.); from 2 to 3 = 4 in. (10 c.m.); from 3 to 4 = 1½ in. (3·75 c.m.); from 5 to 1 = 1¼ in. (3 c.m.).

These troughs can be made by any tinsmith. They should not be used for collecting sulphuretted hydrogen, as this gas will discolour the inside. Water should not be left standing in them when they are not in use, and if they are drained dry each time after use they may be preserved in good condition for many years.

10. To show the inflammability of hydrogen. The gas may be ignited as it issues from a jet. An iron 'fish-tail' burner, which may be screwed into a piece of lead or 'compo' pipe, is well adapted for this, as the flame of the burning gas will not become coloured by foreign substances. If the gas issuing from a generating bottle is used, care must be taken to ascertain that no air is left in the apparatus when the light is applied to the jet; it is preferable to employ hydrogen previously collected in a gas-holder, or contained in a steel gas bottle under pressure.

A lighted taper may be applied to a mouth of a cylinder of hydrogen from which the cover has been withdrawn.

11. To show the lightness of hydrogen. Hydrogen may be poured upwards from one cylinder to another. On introducing a lighted taper into the upper cylinder it can be demonstrated that hydrogen has entered it, and by plunging the taper into the cylinder which previously contained the gas it will be seen to have become entirely free from it.

12. Two cylinders of hydrogen, one held mouth up and the other mouth down, may have their covers removed, and a lighted taper applied. In the first case the up-rushing hydrogen is rapidly all burnt, while in the second the combustion is slowly continuous at the mouth of the vessel. On thrusting the taper up into the gas, the flame of the taper will be extinguished, but it may be re-ignited as it is once more withdrawn through the flame of the burning hydrogen.

FIG. 7.

For this experiment a taper which is attached to a wire should be used, as the heat of the hydrogen flame will soften the wax of an ordinary taper so that it will lose all rigidity. A short piece of thick taper is fastened into a helix at the end of a copper wire, by slightly warming the latter.

13. Hydrogen may be poured upward from a cylinder into a beaker suspended mouth downward from the arm of a balance, and counterpoised. The hydrogen as it ascends into the beaker will cause that end of the beam to rise.

A balance suitable for this and a number of other lecture experiments may be readily constructed in the laboratory :—

(1) *The beam.* Four square rods of wood, A, B, C, D, about 50 centimetres long and 1 centimetre thick, are arranged in the form of an elongated parallelogram, the short diameter of which is about 13 centimetres. The rods are fixed in this position by glueing and nailing thin pieces of wood, E, about 5 centimetres wide, across the short diameter, and small pieces at the two acute angles. As the two thin pieces at the centre would not be able to firmly carry the 'knife-edge,' a block of wood should be glued between them. The 'knife-edge' may be made by grinding one of the edges of a 'three-corner' file upon a grindstone; a piece of the file about 6 centimetres long will be required. A hole is bored in the centre of the beam, and the file firmly hammered into its place. Through each end of the beam is forced a short piece of a steel knitting-needle or a French nail, upon which to hang the hooks intended to carry the pans. The exact position of these two wires should be such that they are

equidistant from the centre of the beam, and that a thread stretched from one to the other will just touch the knife-edge. To secure that the centre of gravity is below the knife-edge, a

FIG. 8.

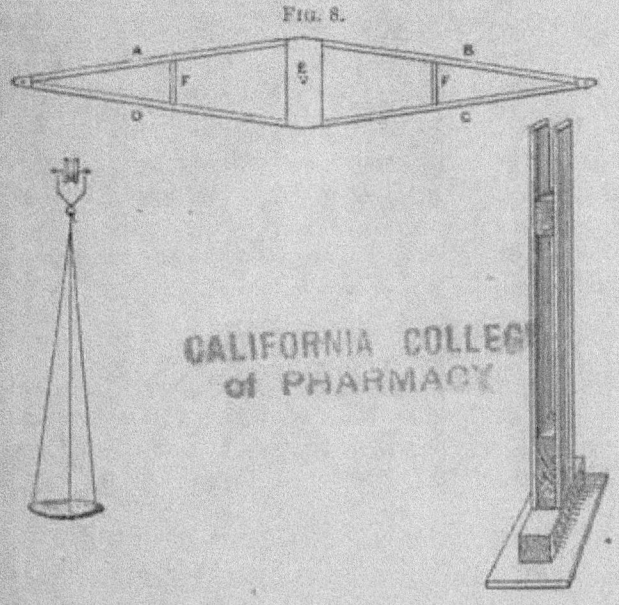

small strip of lead may be attached to the middle of the lower edge of the beam. If desired the beam may be strengthened by fixing cross-pieces of wood at F, F.

(2) *The stand.* Two pieces of wood about 60 centimetres long, 5 centimetres wide, and 6 millimetres thick, are placed parallel to each other at such a distance apart that the beam can freely move between them. They are fixed in this position by being screwed to two blocks of wood, one at one end, the other at some distance from the opposite end. The whole is then fastened to a foot, which may either be sufficiently large and heavy to give stability to the stand, or, better still, bored with a couple of holes so that it can be screwed to the table. Along the top ends of the two uprights, pieces of glass tube are

cemented; upon these the knife-edge will rest, and move with very slight friction.

(3) *The pans.* These may be made of ordinary sheet-iron sand baths, into which three holes may be punched at equal distances round the edge; they may be supported by thin copper wires of such a length that the pans will ultimately hang at about 15 centimetres from the table.

A cardboard scale may be attached to the stand near the foot, and a long straw pointer (made by slipping one straw into another) attached by means of a cork to one end of the file in the beam.

A balance so made, although rough-looking, will be a fairly delicate instrument, and will readily show the one-hundredth of a gram.

14. Soap bubbles may be blown with hydrogen. Hydrogen direct from a generating vessel may be employed, provided the gas is passed through a tube about ten centimetres long, plugged with cotton wool. A thistle funnel is a convenient apparatus upon which to blow the bubbles. The funnel is dipped into the soap solution, and almost as soon as the bubble begins to form it should be directed upwards. As the bubble increases in size it will pull itself off the tube, and as this point is reached, a slight movement of the tube will easily disengage the bubble, which then rapidly ascends.

Much larger bubbles and a more beautiful effect are produced by first blowing a soap bubble in the ordinary way, from the lungs, and by means of a T tube, allowing just enough hydrogen to pass in to enable the bubble to rise slowly through the air when freed from the tube.

Of the many formulæ for preparing soap solutions, the following will be found to give excellent results.

Ten grams of sodium oleate and 400 c.c. distilled water are placed in a stoppered bottle, and allowed to stand until the oleate has dissolved, without warming: 100 c.c. of pure glycerine are then added, and the mixture after being well shaken is allowed to stand, in the dark, for a few days. The clear solution is then carefully decanted, or siphoned, into a clean stoppered bottle, and one drop of strong ammonia solution added. If kept in the dark, and not exposed to the air, this solution may be preserved for years.

15. **Balloons may be filled with hydrogen.** Balloons made of either gold-beater's skin or of collodion may be used. In either case, a short piece of glass tube with smoothed edges is inserted into the neck of the balloon, and a stream of hydrogen from a reservoir gently driven in. As soon as the balloon is full it may be slipped off the glass tube, and the neck secured by twisting between the thumb and fingers. If a collodion balloon be used, it and the contained gas may be set fire to as it ascends through the air. To effect this, the neck of the balloon is tied with a piece of touch-cotton (made by dipping white darning cotton into a solution of potassium nitrate, and drying it) and the end of the cotton ignited by applying to it a piece of burning touch-cotton. A flame, as of a burning taper, must not be used, as it would at once inflame the balloon. To secure that the touch-cotton shall ignite the balloon, a small piece of gun-cotton may be attached to the thread just where it secures the neck of the balloon; the burning touch-cotton will never fail to fire this, and consequently the balloon.

Collodion balloons may be made in the following way: a pear-shaped round-bottomed flask is selected, and carefully cleaned and dried. The reason for employing a flask of this shape is to facilitate the withdrawal of the balloon, which would be an extremely difficult operation if the flask had sharp shoulders. A quantity of collodion (what is known as 'enamelling' collodion answers best) is poured into the flask, and the whole of the inside surface wetted with it by twisting the flask about, the excess being poured back into the bottle. The flask is then held in a support with its mouth downward, so that the last portions of the liquid can drain out and the film gradually set. A current of air may be gently blown into the flask by means of the blowpipe bellows to hurry the evaporation of the ether and alcohol. In this way a very thin film of collodion is obtained adhering to the inside of the flask, and it is quite possible, although somewhat difficult, to remove it without tearing it. The film may be greatly strengthened, however, by the following plan. After the collodion has partially dried, but before that adhering to the lip of the flask has begun to harden and part from the glass, a quantity of a dilute solution of indiarubber in benzol is poured in, and run all over the inside of the flask, care being taken that the whole surface is wetted with it. The excess is run out, and the flask again held up to drain, and a pretty rapid stream of air blown into it. The flask may be gently warmed while this drying process is carried on. (This indiarubber solution is made by placing a few scraps of indiarubber in a bottle, covering them with benzol and allowing it to remain for a few days. A quantity of the clear solution is then poured off and diluted with about ten times its volume of benzol.) It is absolutely necessary to the final removal of the balloon from the flask that none of this solution gets between the collodion film and the glass; hence care must be taken to avoid peeling the thick edges of collodion from the mouth of the flask.

Having now varnished, as it were, the film within the flask, another layer of collodion can be put in, in precisely the same way as the first; in this way a film is obtained consisting of two layers of collodion, cemented together by the varnish of rubber; this film is exceedingly tough and strong. The flask should be put aside for the film to dry spontaneously. In a few hours the film will be seen to have parted from the glass, owing to shrinking on drying. To remove the balloon it is only necessary to insert a glass tube with well-rounded edges, gather the neck together with the fingers, and suck out the air from the inside; this causes the balloon to collapse, and the sloping shoulders of the flask allow of its easy withdrawal.

If the second coating of collodion be made with collodion which is coloured—e.g. by adding to it a small quantity of an alcoholic solution of some aniline colour, as magenta—not only is a coloured balloon obtained, but it makes it easy to see that the second coat of collodion covers the entire surface, when it is being poured round the inside of the flask.

Sometimes, owing to some slight roughness upon the inside surface of the flask, it is impossible to remove the balloon without tearing it; in this case another flask must be taken, and when one has been found which answers the purpose, it should be reserved for this use alone, being carefully corked up before being put away.

16. The lightness of hydrogen may be demonstrated by the shadow which a stream of the gas will cast upon a white surface, which is illuminated by a strong light. The light from an electric lamp is caused to fall upon a screen, without the use of any lenses; a large glass vessel filled with hydrogen is held at a few feet from the screen, so as to cast a shadow, with its mouth downward. On removing the cover and pouring the gas up out of the vessel, a shadow of the uprising stream of hydrogen will be seen owing to the difference in the refractive power of hydrogen and air. Instead of pouring the gas out of a vessel, a stream of it from a reservoir may be made to issue from a wide glass tube—a retort adaptor answers very well. The mouth of the tube should be directed as shown

FIG. 9.

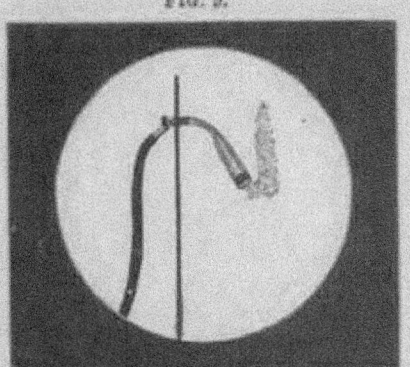

in Fig. 9, in order to give the hydrogen an initial downward motion, and by the shadow it will be seen to curl up. To prevent the scattering of light from the edges of the glass tube, the end of it may be conveniently covered by slipping over it a short piece of wide indiarubber tube, or encasing it in black paper.

17. To show that a mixture of hydrogen with air will explode on ignition. A soda-water bottle (a pear-shaped bottle should be used, as being stronger than those with angles) is fitted with a cork, through which passes a short piece of clay tobacco-pipe stem, projecting about a centimetre beyond the cork. A small quantity of granulated zinc is placed in the bottle and just covered with water, and a few drops of strong sulphuric acid added; the cork is then instantly replaced, and a lighted taper held to the end of the piece of tobacco pipe. In a few seconds the mixture of air and hydrogen will explode, driving the cork with some violence from the mouth of the bottle. Care should be taken to hold the bottle in such a position that the cork will be shot where it can do no damage.

18. A tubulated bell-jar is fitted with a cork carrying a short straight piece of glass tube, which is closed by a small caoutchouc stopper. The bell-jar is stood in a shallow dish (a soup plate) of water, and filled with hydrogen by displacement; a stream of the gas being admitted through the tube at the top, and the displaced air being allowed to bubble out through the water in the plate. When the jar is full, the glass tube is closed by means of the caoutchouc stopper. On raising the bell-jar, and removing the small stopper from the glass tube, and applying a light to the orifice, the hydrogen will burn quietly as it issues from the tube (fig. 10). The air which enters the bell-jar from below, as the hydrogen is making its escape from the top, will soon form an explosive mixture, and in a few moments the flame will strike back into the bell-jar with a detonation. This experiment illustrates at one and the same time the lightness of hydrogen, its inflammability, and the explosibility of its mixture with air.

FIG. 10.

19. To show that a mixture of hydrogen and oxygen will explode on ignition. A soda-water bottle is filled with a

mixture of two volumes of hydrogen and one volume of oxygen, roughly measured. On applying a lighted taper to the mouth of the bottle, a loud explosion will result. Although it is extremely rare for one of these bottles to break with the force of the explosion, it is advisable to envelop it in a towel before igniting the gases.

20. A collodion balloon may be filled with a mixture of hydrogen and oxygen, secured with touch-cotton (as described in No. 15), and allowed to ascend; when the burning touch-cotton ignites the balloon, the gaseous mixture will explode. Rather an excess of hydrogen should be present in the mixture, both to secure the ascent of the balloon and to make up for the greater rapidity with which this gas escapes by diffusion than the oxygen.

21. A froth of soap-bubbles may be blown by sending a stream of the mixed gases through some soap solution contained in a mortar. On applying a flame to the froth a loud explosion ensues. Care must be taken to remove the tube which delivers the mixed gases into the soap solution before igniting the mixture.

A small quantity of the soap solution may be poured into the hollow of the hand, and a froth blown upon it in the same way. On applying a lighted taper the gas in the bubbles explodes with a sharp report, but the explosion is quite unfelt by the hand.

22. To show a singing flame. A flame of hydrogen is burnt from a fine metal jet (the jet from an ordinary mouth blowpipe attached to a straight tube), and glass tubes of various lengths are lowered over the flame; when the right relation between the size of the flame and the draught up the tube is established the flame begins to sing.

23. To show the condition of the flame while singing. Hydrogen which is charged with sufficient naphtha to make it burn with a luminous flame is made to give a singing flame. An image of the flame is projected by means of a lens upon

a piece of looking-glass, and reflected thence upon a screen. If, while the flame is quiet (which may be effected by diminishing the draught up the tube, by placing a small piece of card or mica across the top), the mirror be moved so as to cause the image of the flame to oscillate, a continuous streak of light will be seen; on again allowing the flame to sing and continuing the motion of the mirror, the streak will appear broken up into a number of flashes, or a succession of images of the flame, the musical note being produced by the rapid succession of small explosions, caused by the tendency of the draught up the tube to lift the hydrogen flame away from the jet.

Hydrogen may be naphthalised by being passed through a small Woulf's bottle, containing cotton wool or tow, moistened with naphtha.

24. **To show the product of the combustion of hydrogen in air.** A clean and dry glass bell-jar, held for a few moments

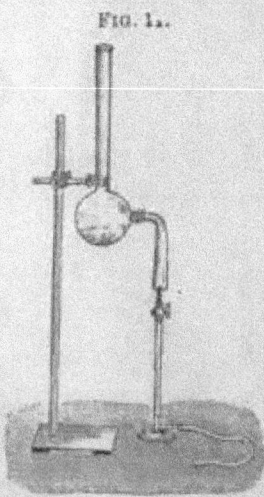

FIG. 1..

over a burning jet of hydrogen, will be instantly bedewed with moisture. A considerable quantity of water may be collected by burning the hydrogen beneath the end of a wide glass tube, bent at right angles, and with its other end fitted into the tubulure of a receiver or a retort. The draught up the tube may be diminished by partially closing the upper and open end of the apparatus, either by means of a strip of mica, or by inserting a funnel; if the draught is too rapid, the receiver will get warm and the water vapour resulting from the combustion will be carried away. When sufficient water is collected, it may be poured out into a test glass, and tested by dropping into it a small fragment of potassium.

25. To show the combination of hydrogen with oxygen by means of spongy platinum. A little bundle of asbestos, attached to the end of a copper wire, is dipped in a solution of platinum chloride, and heated in a Bunsen flame. If this platinised asbestos be held in a stream of hydrogen gas, it becomes red hot, and will ignite the hydrogen.

26. To show the colour of incandescent hydrogen. When electric sparks from a Ruhmkorf coil are passed through hydrogen in a tube, the light emitted is of a feeble blue colour, but if the sparks are intensified by means of a Leyden jar, or other condenser, the colour of the glowing hydrogen will be seen to be a brilliant rosy red. If the gas is under reduced pressure, the length of the spark which a coil is capable of giving will be much increased, the intensity of the light emitted will, however, be diminished. A glass tube about 10 c.m. long, having a short branch tube blown upon it in the middle, has platinum wires fused into the ends, as shown in the figure. The branch tube, which is constricted, is connected to a Sprengel pump, and the tube exhausted. Hydrogen is then admitted, and the pump slowly set in operation until the point of rarefaction is reached, at which the spark will pass between the wires. This is ascertained by frequent trial with the coil during the exhaustion. The tube is then sealed at the constriction.

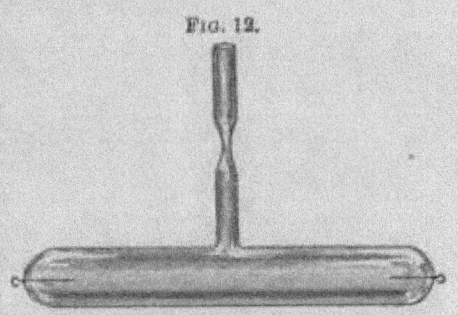

FIG. 12.

27. To show the high conductivity of heat of hydrogen. Two stout copper wires are thrust through a cork, and a loop of thin platinum wire, about two inches long, soldered to them at one end, so that it can be heated by an electric current. The current necessary to raise the wire to incandescence, in the air,

must be arranged, and on thrusting the wire up into an inverted cylinder of hydrogen, the glow will be seen to greatly diminish, or to cease altogether, depending upon the original intensity of the glow when in the air.

FIG. 13.

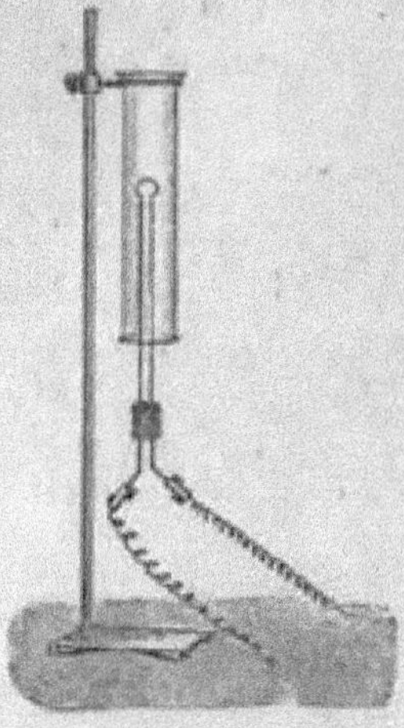

As introduction of the heated wire into the hydrogen causes the ignition of the gas at the mouth of the cylinder, it is best to pass the wire in before connecting with the battery (a switch should be placed in the circuit). When the circuit is completed, no visible effect results, but on withdrawing the wire into the air it will instantly glow, and the withdrawal of the wire will not inflame the hydrogen. This may be repeated once or twice if the air is fairly free from draught, and finally the glowing wire may be thrust in without interrupting the current; the hydrogen will inflame, and the glow will again stop.

HYDROGENIUM

28. The fact that palladium has the power of absorbing or condensing upon itself large quantities of hydrogen may be

FIG. 14.

shown by passing an electric current through acidulated water, using a plate of palladium as the negative electrode. A narrow vertical glass cell may be employed, and an image of it projected upon the screen. When two platinum electrodes are introduced, and a current from three Grove's cells passed through the liquid, bubbles of gas will be seen to be disengaged from both plates; on replacing the negative plate by a similar plate of palladium, and again passing the current, gas will be evolved only from the positive electrode, the hydrogen being entirely absorbed by the palladium. The palladium electrode should be heated to redness immediately before being introduced. After a short time the palladium will have absorbed as much hydrogen as it is capable of taking up, and when this point is reached gas will be seen to be evolved from both electrodes.

29. To show the chemical activity of the hydrogen in this condition by its reducing power. The palladium electrode, after it has absorbed hydrogen, may be immersed for a moment in a solution of ferric chloride, and quickly withdrawn. The solution which adheres to the metal will be more or less reduced by the absorbed hydrogen to the ferrous condition, and if the electrode is then dipped into a dilute solution of potassium ferricyanide (which may be contained in a cell and an image thrown on the screen) the blue colour will instantly be seen.

30. To show the change in volume which palladium undergoes when it absorbs hydrogen. Two narrow strips of palladium, each attached to a platinum wire, are varnished on one side with either shellac or photographic varnish. These strips are then placed in a suitable cell containing water acidulated with sulphuric acid, and an image of them projected upon the screen. On passing the current from two Grove's cells, the negative electrode will be seen to curl up; on reversing the current it will uncurl, and the other one, which has now become the negative plate, will begin to curl; and they will continue to curl and uncurl as often as the current is reversed.

FIG. 15.

The most suitable form of cell for this experiment is one which can be used with a horizontal projection apparatus (see Lantern Illustrations). It consists of a shallow circular, flat-bottomed glass dish, of nearly equal diameter to the condensing

lens of the apparatus, and should be provided with two binding
screws to which the electrodes may be attached (fig. 15). The
electrodes should be of such a length that as they curl and
uncurl they can pass each other without touching. They must
be placed in the cell edgeways to the light, and their varnished
sides should be so arranged that the strips will curl up in
opposite directions.

OXYGEN

METHODS OF PREPARATION

31. From mercuric oxide. $HgO = Hg + O$. A piece of combustion tube closed at one end is bent so as to produce an elbow
at a short distance from the closed end (fig. 16), the open end
is fitted with a cork and delivery tube. A small quantity of
mercuric oxide is introduced into the tube, and strongly heated.
Oxygen is evolved, and may be collected over the pneumatic
trough. Mercury condenses on the sides of the tube at a short
distance from the flame, and the globules will collect at the
bend.

FIG. 16.   FIG. 17

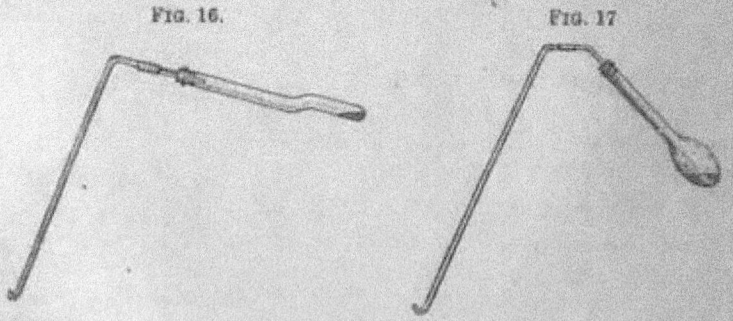

32. From potassium chlorate. $KClO_3 = KCl + 3O$. Potassium chlorate may be heated in a flask (an assay bulb is very

convenient), fitted with a cork and delivery tube (fig. 17). It is well to previously dry the salt, otherwise moisture from it condensing on the neck of the flask is liable to run back upon the hot glass and cause its fracture.

33. From a mixture of potassium chlorate and manganese dioxide. Chlorate of potash (not powdered) may be mixed with about its own bulk of manganese dioxide in a Florence flask. The flask should be held in a horizontal position, as a considerable quantity of moisture is disengaged from the mixture, and in this way all risk of its condensing and running back into the hot flask is avoided. As the evolution of oxygen is very rapid, wide delivery tubes must be used; it is best to stretch a piece of caoutchouc tube over the neck of the flask (as shown in fig. 18). If the gas is to be washed, which it is desirable to do,

FIG. 18.

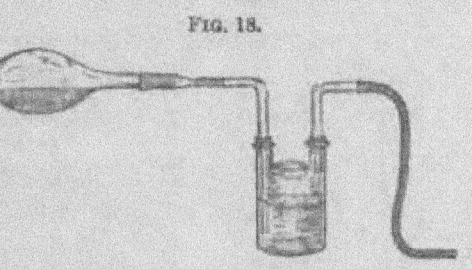

as considerable quantities of oxide of manganese are carried forward with the gas, glass tubes should be fitted to a Woulf's bottle, as wide as the necks will allow. The gas may be collected in a gasholder.

34. By the action of chlorine upon steam at a red heat. $H_2O + Cl_2 = 2HCl + O$. A stream of chlorine gas is passed through a flask containing water, and kept at the boiling temperature (F, fig. 19).[1] The mixture of chlorine and water vapour is then conducted into a porcelain tube, filled with fragments of pumice, and heated to redness in a furnace. Immediately before entering the hot tube the gases should be passed through an empty flask or Woulf's bottle, to arrest condensed water, and, as an additional safeguard, the lower end of the

[1] For the description of the chlorine-generating apparatus see Exp. 151.

tube leading to the porcelain tube may be filed off obliquely (as shown in fig. 20) so that any drops of water condensing in this

FIG. 19.

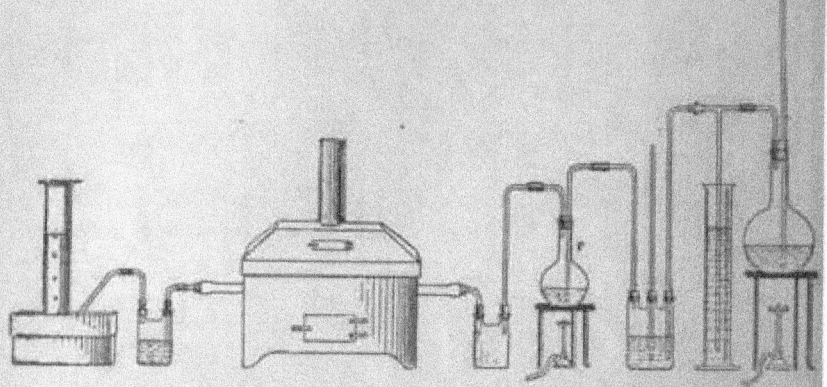

tube may be able to run back into the bottle.   The gases which are evolved at the end of the heated tube must be passed through a Woulf's bottle containing a solution of sodium hydroxide to absorb the hydrochloric acid which is formed in the reaction, and the oxygen may be collected over the pneumatic trough.   It is desirable to encase the porcelain tube in a thin sheet-iron tube, which makes it less liable to fracture.

FIG. 20.

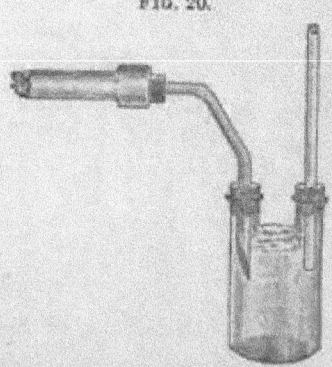

35. By the action of chlorine and cobaltic oxide on water.

$$(1)\ Cl_2 + H_2O + 2CoO = 2HCl + Co_2O_3 ;$$
$$(2)\qquad\qquad Co_2O_3 = 2CoO + O ;$$
$$(3)\quad 2HCl + 2NaHO = 2NaCl + 2H_2O ;$$

or,

(1) $2NaHO + Cl_2 + 2CoO = 2NaCl + H_2O + Co_2O_3$;

(2) $\qquad\qquad Co_2O_3 = 2CoO + O.$

A few drops of a solution of a cobalt salt (chloride or nitrate) are poured into a flask of about 400 c.c. capacity containing a strong solution of caustic soda. This solution is gently boiled and a stream of chlorine passed through the liquid, when a rapid evolution of oxygen will take place. The tube which delivers the chlorine should dip about a centimetre beneath the liquid, and should be made as wide as can be conveniently passed through the neck of the flask to prevent a stoppage by the sodium chloride which is formed; the wide piece may be blown on to a narrow tube as shown in the figure, or a thistle funnel may be substituted.

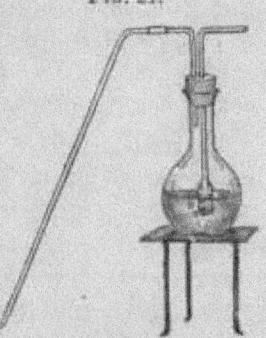

Fig. 21.

36. By the action of cobaltic oxide on a solution of bleaching powder.

(1) $Ca(OCl)_2 + 4CoO = CaCl_2 + 2Co_2O_3$;

(2) $\qquad 2Co_2O_3 = 4CoO + O_2.$

A few drops of nitrate of cobalt solution are added to a solution of bleaching powder contained in a flask (about 250 c.c. capacity) fitted with a cork and bent delivery tube; on gently heating the mixture oxygen is evolved.

37. By passing a stream of chlorine gas through boiling milk of lime to which a few drops of a solution of a cobalt (or nickel) salt have been added. $Cl_2 + CaH_2O_2 = CaCl_2 + H_2O + O.$ The milk of lime is placed in a flask of about 400 c.c. capacity, and a small quantity of cobalt chloride added. The flask is fitted with a cork and two tubes, one of which passes nearly to the bottom of the flask, the other ending below the cork The con-

tents of the flask are raised to the boiling-point when chlorine is passed through, and a rapid evolution of oxygen takes place.

38. By the action of water upon sodium peroxide. $Na_2O_2 + H_2O = 2NaHO + O$. A few grams of sodium peroxide are placed in a small dry flask, furnished with a dropping funnel and delivery tube, and water is slowly introduced. Each drop of water falling upon the peroxide causes an energetic reaction to take place, with the rapid evolution of oxygen, which may be collected at the pneumatic trough.

A modification of this experiment may be made in the following way. A small heap of sodium peroxide is placed upon a piece of blotting-paper, and a drop of water from a pipette is allowed to fall upon the compound. The energy of the action at once causes the ignition of the paper.

Or again, if a little sodium peroxide be mixed with about its own bulk of magnesium filings, and a few drops of water are added to the mixture on the lid of a porcelain crucible, the mass immediately inflames with a mild detonation.

39. By gently heating a mixture of manganese dioxide and strong sulphuric acid in a flask provided with a delivery tube, oxygen gas is evolved. $MnO_2 + H_2SO_4 = MnSO_4 + H_2O + O$. The manganese dioxide may be replaced by either lead dioxide or red-lead.

40. By gently heating a few crystals of potassium dichromate with strong sulphuric acid in a flask with a delivery tube, oxygen is rapidly evolved. $K_2Cr_2O_7 + 4H_2SO_4 = K_2SO_4 + Cr_2(SO_4)_3 + 4H_2O + 3O$.

41. By the action of plants upon carbon dioxide in bright light. For this purpose sprigs of fresh mint are placed in a glass cylinder filled with water charged with carbon dioxide and inverted in a dish of the same solution. The whole is then exposed to bright sunlight for several hours. Bubbles of gas

gradually make their appearance and collect at the top of the cylinder. The gas may be transferred to a test tube, shaken with a few drops of caustic soda solution and finally shown to be oxygen by its effect on a glowing splint of wood.

42. If the mint be placed in a glass cell filled with the carbonic acid solution, and an image of it thrown on the screen by means of the electric light, the action of the light in bringing about the decomposition will be seen by the rapid formation of bubbles of gas upon the leaves, and in a very few moments bubbles of oxygen will be seen to detach themselves from the leaves and float to the surface. The beam of light must be passed through a water cell to cut off the heat before being allowed to enter the cell containing the mint.

43. Another way of showing this effect is to pack a large flask, capable of holding about six litres, with fresh green stuff, such as mint. The flask is closed by a cork carrying two tubes, one reaching to the bottom, the other ending just through the cork. A stream of washed carbon dioxide is then passed through the flask, entering by the longer tube, until the air is entirely displaced, and the escaping gas is entirely absorbed by caustic soda. The delivery tube in connection with the exit tube from the flask may be made to dip into a dish containing a solution of caustic soda. A long tube, having a stoppered funnel at one end, is filled with soda and inverted over the end of the delivery tube, and the bubbles of carbon dioxide which are allowed very slowly to pass through the apparatus will be seen to entirely absorb as they ascend the tube. The flask is then exposed to the light from a powerful electric arc, and almost immediately it will be seen that the ascending bubbles are not entirely absorbed, but leave a small residual quantity of gas. As the upper layers of caustic soda in the tube get saturated with carbon dioxide, fresh quantities of the solution may be allowed to enter from the stoppered funnel at the top of the tube. In a short time several cubic centimetres of oxygen may be collected.

44. To show the action of oxygen on combustibles. A chip of wood which has been lighted and blown out, and still retains a glowing spark upon it, may be introduced into a cylinder of oxygen; vigorous combustion instantly begins, and the chip will burst into flame. The cedar splints retailed by tobacconists are most convenient for this purpose.

The same effect may be obtained by introducing a taper, the wick of which retains a glowing spark upon it. The kind of taper best suited for this is the green-coloured taper, usually obtained in coils. A short piece is fastened into a helix at the end of a copper wire.

45. A piece of charcoal may be attached by means of copper wire to a deflagrating spoon, a corner of it ignited in a flame, and the charcoal introduced into a flask of oxygen.

If bark charcoal be used, it will throw off showers of scintillations as it burns. A supply of bark charcoal can easily be obtained by packing a crucible with pieces of bark, covering the crucible, and heating it to redness for some time in a fire.

46. A piece of sulphur contained in a deflagrating spoon may be ignited and introduced into oxygen.

Instead of placing the sulphur in the spoon, the spoon itself may be unscrewed or cut off, and in its place a bundle of asbestos is fastened, by means of copper binding wire; a quantity of sulphur is melted in a test-tube, and the asbestos dipped into it. In this way a large surface of sulphur will be obtained, and when it is ignited and plunged into oxygen a flame 6 or 7 centimetres in length will be obtained. The same bundle of asbestos can be again dipped in sulphur and used for an indefinite number of times.

47. A piece of phosphorus may be ignited on a deflagrating spoon and introduced into a flask of oxygen. The spoon should reach nearly to the bottom of the flask (the cap being fixed at the right place upon the stem before the experiment is made), so that the flame of the burning phosphorus may not impinge against the upper part of the flask, and so crack it.

48. To show the action of oxygen on certain metals. A bundle of steel wire may be burnt in oxygen by directing a stream of the gas through a spirit-lamp flame, and causing the flame to impinge upon the end of the bundle of wire, the wire being held in the same direction as the flame (fig. 22). The wire will thus become ignited, and the lamp may be withdrawn. The bundle of wire continues to burn with brilliant scintillations in the stream of oxygen. The oxygen may be delivered through

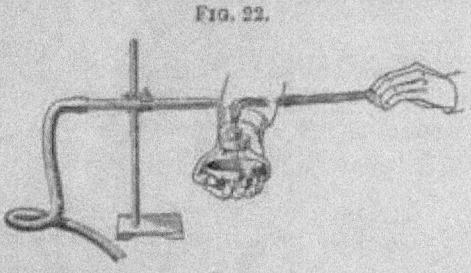

Fig. 22.

a piece of glass tube, about 5 millimetres bore.

Instead of a spirit-lamp flame a Herapath blowpipe may be employed, using coal gas for the flame with which to heat up the wire, and in the same way the gas may be turned off as soon as the steel begins to burn.

49. A stream of oxygen may be passed through molten iron in a crucible. A quantity of scrap cast iron is melted in a Hessian crucible in a suitable furnace. A stream of oxygen, delivered through a piece of iron gas pipe, is forced through the metal, and a brilliant shower of sparks will be thrown up out of the crucible. The end of the gas pipe must be made red hot before being introduced into the crucible, or it will be instantly stopped up.

50. Powdered magnesium may be burnt in oxygen by injecting a small quantity of the metal into a flame by means of a sudden puff of oxygen gas.

For this purpose a piece of brass tube, about 10 centimetres long and 10 millimetres bore, is connected at one end by means of a caoutchouc tube to an oxygen reservoir. A small quantity (2 or 3 grams) of magnesium filings is placed in the brass tube at the open end. A roll of wire gauze, about 6 centimetres long and about the same thickness as the brass tube, is threaded upon a piece of copper wire, the end of which is wound into a spiral through which the brass tube can just pass ; the roll of gauze is dipped into spirit, and the brass tube pushed into the spiral until the open end nearly touches the gauze; the spirit is then inflamed. The air contained in the caoutchouc tube is very gently swept out by passing a slow stream of oxygen through the apparatus, and then on sending a sudden puff of gas the magnesium will be projected into the long spirit flame, and at the same time supplied with oxygen, so that the flash of light will be intensely dazzling.

51. Aluminium may be burnt in oxygen by introducing into a cylinder of oxygen two or three leaves of aluminium (books of aluminium leaf are to be obtained, just as gold leaf or Dutch metal from the dealers), and dropping upon them a small fragment of charcoal, about the size of half a pea, which has been ignited in a flame. The metal will burn with a brilliant flash of light.

52. Zinc may be burnt in oxygen by directing a stream of the gas into a crucible, containing a quantity of zinc heated in a furnace until the zinc begins to burn in the air.

53. To show the combustion of hydrogen in oxygen, a jet of hydrogen burning from an iron fish-tail burner, which is screwed into the end of a piece of lead pipe, may be lowered into a cylinder of oxygen.

To show the high temperature of the flame of hydrogen burning in oxygen, hydrogen may be burnt in an ordinary oxy-hydrogen jet, and gently fed with oxygen.

On directing the flame down upon a fragment of metal, such as silver, contained in a bone-ash cupel, the metal may be boiled.

A piece of thin steel—e.g. a fragment of watch-spring—held in the flame burns with the evolution of brilliant scintillations.

A piece of platinum wire held in the flame rapidly melts, and the molten globule begins to boil on the end of the wire.

A fragment of lime presented to the flame is instantly raised to incandescence, and emits a brilliant light, well known as the 'lime light' or 'Drummond light.'

54. To show the solubility of oxygen in molten silver. A quantity of silver is heated, as described in Exp. 53, until a large button of the molten metal is obtained; the supply of hydrogen is gradually diminished, so that there is an excess of oxygen passing from the jet. The melted metal will rapidly absorb all the oxygen it is capable of retaining. On allowing the mass to cool, this oxygen is again evolved, and as the metal solidifies first on the outside, the crust will be violently burst by the sudden evolution of oxygen from the interior portions of the mass.

OZONE

METHODS OF PREPARATION

55. By the slow oxidation of phosphorus. One or two sticks of freshly scraped phosphorus are placed at the bottom of a tall glass cylinder and about half covered with water. Within five or ten minutes enough ozone will be produced to answer the test with ozone paper. The cylinder may be closed either by a stopper or by a glass plate. After the lapse of some time the ozone which is formed begins to act upon the phosphorus, and so diminish in quantity, until at last there will be insufficient left to give the test; it is necessary, therefore, to avoid exposing the phosphorus too long before the experiment is required.

56. By the slow oxidation of ether. A few cubic centimetres of ether are poured into a glass cylinder, and a spiral of platinum wire, which has been heated by passing through a Bunsen flame, is lowered to within two or three centimetres of the ether. The

mixture of ether vapour and air undergoes combustion upon the platinum wire, and if a piece of moistened ozone paper be introduced into the cylinder the blue colour will instantly make its appearance. (Whether this is really due to ozone is a question which has not yet been put beyond doubt.)

Ozone papers (syn. 'starch papers') may be made by dipping strips of cartridge paper into an emulsion of starch to which a small quantity of a solution of potassium iodide has been added, and hanging them up to dry; they may be preserved in stoppered bottles. When used they are first moistened by being dipped into distilled water.

57. **By the passage of electric sparks through the air.** The air in the immediate neighbourhood of an electrical machine at work is so charged with ozone that a moistened test paper suspended near such a machine is very soon coloured. If the paper is held between the terminals, especially when they are giving a brush discharge, the paper is instantly turned blue.

FIG. 23.   58. **By the passage of an electric current through acidulated water.** A current from five or ten Grove's cells may be passed through water acidulated with sulphuric acid, using the apparatus designed for collecting the mixed gases. On allowing the escaping gases to impinge upon starch paper, the presence of ozone will be made manifest.

To show that the ozone is formed at the positive electrode only, the apparatus fig. 23 may be used, and a small quantity of starch emulsion added to the acidulated water. Immediately before the experiment is made a few drops of potassium iodide solution are added, and the mixture introduced into the apparatus. On passing the current it will be seen that the contents of that limb of the apparatus containing the positive electrode will be turned blue. The mixture of starch and potassium iodide with dilute sulphuric acid will slowly give the blue colour when left to stand; hence the necessity for not mixing them until the time of performing the experiment.

59. By the passage of the silent electric discharge through oxygen. This may be effected by means of a 'Siemens ozone tube' (fig. 24). The stream of oxygen should be dried by being

FIG. 24.

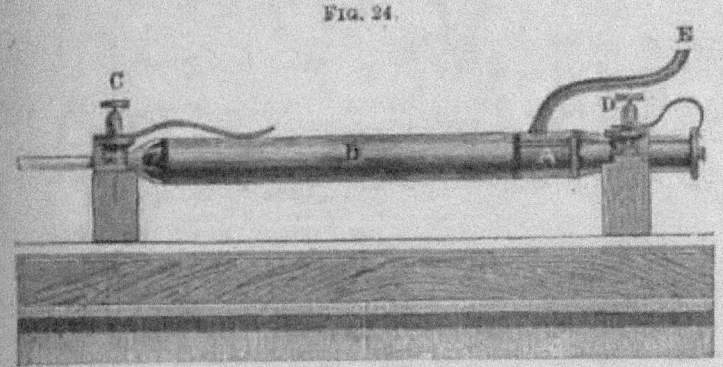

bubbled through concentrated sulphuric acid, and it should be exposed to a feeble electric discharge, obtained by the influence of two Grove's cells upon a Ruhmkorff coil.

60. The Siemens tube may be replaced by a simple piece of apparatus, shown in fig. 25, which may be made in the follow-

FIG. 25.

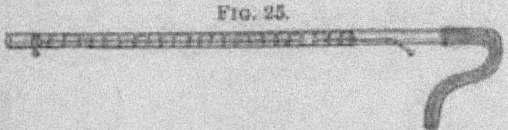

ing way: a piece of glass tube about 30 centimetres long and 4 millimetres bore has a piece of fine platinum wire fused through the side at about 30 millimetres from one end, and extending along the inside of the tube nearly to the other end. A second platinum wire is wound as a spiral round the outside of the tube over its entire length, but ending a few millimetres short of the point where the inner wire comes through the glass, and it is secured from slipping by being tightly twisted round the glass at each end of the long spiral. On sending a gentle stream of oxygen through this tube, the two wires being connected with the coil in action, abundance of ozone is instantly produced.

The simplest way of inserting the platinum through the glass is to heat the part where it is intended to introduce the wire with a fine pointed blow-pipe jet until a small spot is soft; the end of a piece of platinum wire which is heated in the same

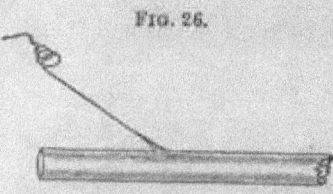

FIG. 26.

flame is then pushed gently into the softened glass, and the glass drawn out by gently pulling the wire (fig. 26); it may be drawn out at any desired angle to the tube. The small branch tube so obtained is cut off near to the main tube, and the long platinum wire pushed through. By directing the blow-pipe flame upon the opening where the wire enters, the glass may be fused to the wire, making a perfectly gas-tight joint.

61. A modification of the Siemens tube, which yields better results than either of the above, may be constructed as follows: A piece of glass tube about 20 centimetres (8 inches) long and 9 millimetres (⅜ inch) bore has one end drawn out and a short

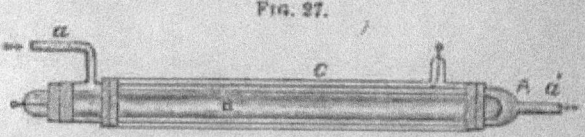

FIG. 27.

piece of small tube blown on (A, fig. 27). As near to the other end as possible a small side tube is blown into the tube.

A second piece of glass tube of the same length, with thin walls, and of such a diameter that it will go inside the other tube, leaving a space of 2 millimetres all round, is sealed up at one end, and a piece of platinum wire about 2 centimetres long fused in at the closed end. The other end is drawn out to a point. The tube is filled with dilute sulphuric acid (1 acid to 2 water), and the end sealed up, leaving only a small air bubble. This tube is to be fixed inside the former one. This may be done by a plug of asbestos paper, a narrow strip being flaked off a piece of asbestos card; this is wrapped round the smaller tube until it fits tightly into the outer one. To keep the inner tube central a similar narrow strip of the asbestos paper may be put round it near the end which is inside, but not so tightly as to prevent the free passage of gas. The outer of these two tubes has now to be surrounded by a third tube, in order that a layer of sulphuric acid may be spread all over the surface. This outermost jacket, into which, near one end, a narrow side tube is blown, may be fitted over the other by means of two narrow rings cut from a piece of caoutchouc tube, and over which the tube will firmly fit. The space between the jacket and the middle tube is filled up with dilute sulphuric acid, and the little branch tube may be drawn off and a platinum wire fused in; or it can be capped with a short piece of caoutchouc tube and a glass stopper carrying a piece of platinum wire.

It will be seen that this piece of apparatus differs from the Siemens tube only in having the two coatings of tin foil replaced by coatings of dilute sulphuric acid, and it is this which increases its efficacy.

62. Ozone appears to be produced in appreciable quantities in all the ordinary reactions for the formation of oxygen; thus when potassium chlorate, or mercuric oxide, is heated, and the gas evolved allowed to impinge upon starch paper, a blue colouration will be produced.

When manganese dioxide or barium peroxide is acted on by sulphuric acid; in the same way ozone is produced.

63. To show that contraction takes place when oxygen is converted into ozone. A piece of apparatus somewhat similar to

the ozone tube above described, but without the outer jacket, may be used for this purpose. In this case, however, the side tube (*a*, fig. 27), has a stopcock upon it, and the tube *a'* is continued for about 30 centimetres. This tube is bent in the

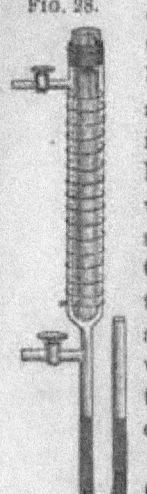

FIG. 28.

middle, so as to form a U tube, to serve the purpose of a manometer. Just beyond the point where this tube is joined to the wider one, a second stopcock is blown in (fig. 28). By means of these stopcocks the apparatus may be filled with oxygen, without disturbing the liquid in the manometer. The U tube may be half filled with sulphuric acid, which may be coloured with indigo to make it more visible. Round the outside of the wider tube a platinum wire is wound. On connecting this with one terminal of an induction coil, and the inside tube of acid to the other, and passing the electric discharge, a contraction in volume of the contained gas will at once be seen by the change in level of the liquid in the two limbs of the manometer.

For this experiment the ozone tube itself (either the Siemens tube or one of the modifications above described) may be used, by attaching to the end of it a narrow U tube containing coloured sulphuric acid. The attachment must be made by means of a short piece of wide

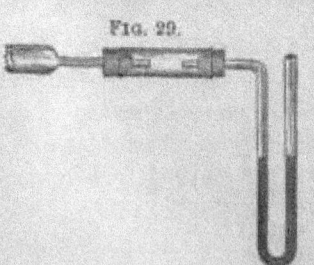

FIG. 29.

tube, fitted with two wooden corks, as shown in fig. 29. A stream of oxygen is first passed through the ozone tube, the manometer is then attached, and the inlet tube stopped by means of a plug of wax. This arrangement, however, does not readily allow of the levels of the liquid in the manometer being adjusted for any change of temperature which may occur between filling the apparatus and performing the experiment.

64. **To show the absorption of ozone by turpentine.** By means of the apparatus, fig. 29A, the contraction in volume which accompanies the conversion of oxygen into ozone may be first demonstrated, and afterwards the further diminution in volume resulting from the absorption of the ozone from the mixture by means of turpentine.

FIG. 29A.

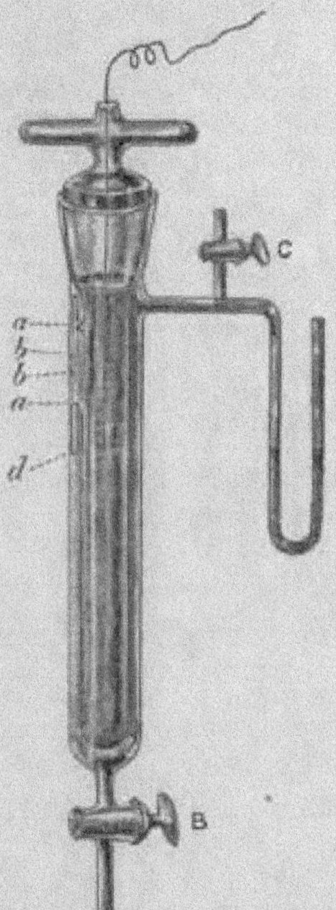

In this apparatus the inner tube is practically a greatly elongated hollow stopper, ground to fit the neck of the outer tube. Near its upper end it has two little projections, a, a, about $2\frac{1}{2}$ centimetres apart.

The outside tube has a small manometer attached, of rather narrow tube, and of such a size that the whole of it will be covered by the lamp condenser, so that an image of it can be thrown upon the screen. The stopcocks B and C allow of a stream of oxygen being passed through the apparatus.

The outer tube has also two little projections upon it, b, b, very near together in a horizontal plane, and in such a position that they lie about midway between those on the inner tube. By means of these four projections, a sealed capillary tube, d, containing turpentine, can be held in the position shown in the figure, and when desired it can be broken by a slight twist of the stopper, without in any way disturbing the level of the liquid in the gauge.

To perform the experiment, the outer tube is held in an inclined position, and a sealed capillary tube filled with turpentine is placed inside, so that it lies between the two projections. The stopper is then inserted and gently turned round until the capillary tube is securely held by the projecting points. Oxygen is then admitted by the cock c, and allowed to escape through B, after which both cocks are closed.

The apparatus, held by its neck in a clamp, is then placed in a jar containing water and crushed ice, the gauge projecting outside, so that an image of it can be thrown upon the screen. As the tube cools, the liquid in the manometer (strong sulphuric acid) should be levelled by admitting a little more oxygen through c.

By means of a wire dipping into the water in the jar, and another into the dilute acid in the inner tube, the oxygen within the apparatus can be ozonised, and the contraction will be manifest by a change of level in gauge. As soon as a moderate contraction has taken place the electrification should be stopped. Now by turning the stopper the capillary tube will be broken, without any disturbance of the image upon the screen, and the turpentine set at liberty, when a further diminution of the gaseous volume will at once take place, which will be indicated by a further movement of the liquid in the manometer.

This second contraction will be seen to be exactly twice as great as that which first resulted.

65. To show the decomposition of ozone by potassium iodide, and no change in the volume. A capillary tube is filled with an aqueous solution of potassium iodide, and introduced into the ozone tube in the manner described above. The experiment is conducted in the same way as when turpentine is used, except that ozonisation may be continued until there is a rather greater contraction in volume. When the little tube is broken no alteration in volume is suffered by the gas, but on removing the tube from the ice, the liberated iodine will be evident by the red colour of the liquid which has been thrown out of the capillary tube.

66. To show the decomposition of ozone by heat. A piece of small glass tube, about 18 centimetres long (7 inches), is attached to the end of the ozone tube by means of the connector above mentioned (the ordinary method of connecting by means of caoutchouc not being available on account of the action of ozone upon it). A slow stream of oxygen is driven through (the rate being observed by the bubbling of the gas through the drying bottle)[1] and the coil set in action. The ozone, which may be shown to be formed, will be entirely destroyed by gently heating the tube by passing a Bunsen flame along it. Unless the stream of gas be carefully regulated, some of the ozone may get through the hot tube without being decomposed, and a test paper held to the orifice of the tube will not show the absence of ozone.

FIG 30.

67. To show the decomposition of ozone by certain metallic oxides. A similar glass tube to the above may be partly filled with copper oxide, and attached to the ozone tube. As the ozonised oxygen passes through, every trace of ozone will be decomposed, and the escaping gas will have no action on starch papers.

68. To show the bleaching action of ozone. A stream of ozonised oxygen is delivered into a tall stoppered cylinder, and a quantity of a dilute solution of a colouring material, such as aniline blue, carmine, or indigo, is poured in and agitated with the gas in the cylinder. If the colour is not discharged at once, more ozone may be passed in, and the mixture shaken again.

[1] In all these experiments it is desirable to bubble the oxygen used through sulphuric acid. This not only dries the gas, but enables the experimenter to judge of the rapidity with which it is passing. A convenient form of wash bottle for this and other similar purposes is seen in fig. 30. It consists of a short T tube, through which a smaller and longer tube is passed, the one being fitted into the other by a short piece of caoutchouc tube. The wider tube may be fitted into a small bottle or cylinder in the same way, with caoutchouc.

69. **To show the action of ozone upon mercury.** A quantity of ozonised oxygen is delivered into a large stoppered bottle or cylinder, and a few cubic centimetres of clean mercury poured in, and rolled round the glass sides of the bottles.

70. **To test for ozone by means of potassium iodide and litmus.** Sheets of blue litmus paper are dipped in water, which has been rendered very feebly acid by the addition of one or two drops of dilute sulphuric acid, and to which a very small quantity of potassium iodide has been added. The papers, which are to be only just made red, are dried, and may be preserved. On moistening one of these test papers and exposing it to ozone, it is turned blue by the alkali liberated by the decomposition of the potassium iodide. If too much iodide is present, the liberated iodine will mask the blue colour by turning the paper brown.

WATER. H_2O

SYNTHETICAL FORMATION

71. **By the combustion of hydrogen in air.** (See Exp. No. 24.)

72. **By the combustion of hydrogen in oxygen.** A piece of brass tube about 10 centimetres long and 9 millimetres bore

FIG. 31.

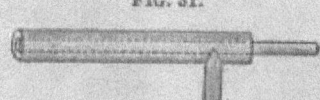

has a small branch tube silver-soldered, or brazed, into it at right angles, near to one end (fig. 31). The end nearest to this branch is closed by a small brass disc, having a hole in it 5 milli-

metres diameter. Through this hole is passed a second brass
tube, which is brazed in place so that the farther end is level
with the end of the wider tube, and the other end projects
about 4 centimetres. A platinum tube about 12 centimetres
long is fitted over the open end of the outer tube by means of a
packing of asbestos paper (see Exp. No. 7). A stream of hydrogen is passed in through the smaller internal tube, and ignited as
it escapes from the end of the platinum tube. Oxygen is then
admitted through the branch tube, and the flame at once strikes
back to the end of the brass jets, and continues to burn inside
the platinum tube, volumes of steam being evolved from the
open end. This steam may be at once condensed by fitting a
glass adaptor into the open end of the platinum tube, again by
means of asbestos, and connecting to this a long piece of glass
tube (fig. 32). A continuous stream of water may in this way

FIG. 32.

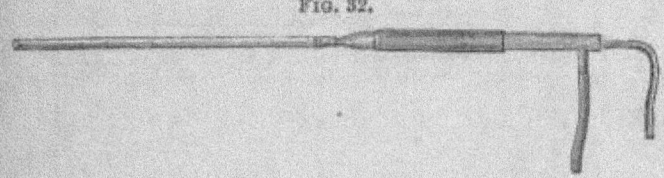

be made to drop from the end of the tube, the rapidity with
which it is formed being only limited by the heat inside the
platinum tube rising high enough to melt the metal.

73. By the explosion of mixed oxygen and hydrogen in a
'Cavendish' eudiometer. A mixture of oxygen and hydrogen,
roughly in the proportion of two volumes of hydrogen to one
volume of oxygen, is made in a glass bell-jar, fitted with a stopcock at the top, and floating in a wide glass jar. (It is convenient
to have this small floating gas-holder roughly graduated.)

The 'Cavendish' eudiometer, after being carefully dried
inside, and having its stoppers greased, is attached to an air-
pump and exhausted; it is then screwed on to the cock of the

floating gas-holder, and the cocks opened; the mixed gases will enter the eudiometer, which is then removed from the bell-jar.

FIG. 33.

The cock of the gas-holder should be opened *before* the one on the eudiometer, to prevent the possibility of air being drawn in, and this second cock should be turned very gradually, so that no drops of water are forced into the eudiometer. The fact that gas has entered, will be evident by the change of level of the water in the vessel containing the bell jar. After the passage of the spark the clean and dry sides of the eudiometer will be covered with condensed moisture. If at the conclusion of the experiment a short tube containing dry calcium chloride be dropped into the eudiometer, and the stopper replaced, the vessel will be found quite dry when again required for use.

74. **To show the volume composition of water synthetically.** A mixture of two volumes of hydrogen and one volume of oxygen is exploded in a eudiometer at such a temperature that the resulting steam is retained in the state of vapour. For this purpose a U-shaped eudiometer is employed (fig. 34), the closed limb of which has platinum wires fused in. The open limb is furnished with a branch stopcock tube near the bend. The closed limb, which is graduated into three equal divisions, is surrounded with a wider tube, fitted over it by means of a cork, in such a way that a stream of vapour of amyl alcohol can be made to pass through, entering at the top, and being conveyed away from the bottom to any convenient condenser. The eudiometer is first filled with mercury, the stopcock is then opened so that the open limb of the apparatus will empty itself

A stream of mixed gas direct from an electrolysis apparatus is delivered into the apparatus by means of a straight piece of small glass tube on the end of which is a short piece of caout-chouc tube; this is thrust down the open limb of the eudiometer (fig. 34A), and beneath the mercury, the caoutchouc tube enabling the gas to pass round the bend of the tube, the displaced mercury flowing out by the stopcock. When enough gas has been introduced, the delivery tube is withdrawn, the stopcock closed, and atmospheric pressure restored by

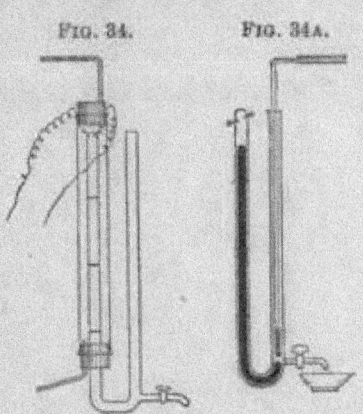

FIG. 34.   FIG. 34A.

pouring mercury into the open limb. The electrolytic gas should be passed through a small drying tube, containing pumice moistened with sulphuric acid. It is well to ascertain for a given piece of apparatus what volume of gas must be delivered into it while cold, in order that it may occupy the space down to the third graduation when the tube is heated by the vapour of amyl alcohol.

FIG. 35.

Immediately before passing the spark, the open limb of the tube must be half filled up with mercury, and the end firmly closed by the thumb. After the explosion, on adjusting the levels of mercury in the two limbs, the gas in the eudiometer will occupy two volumes. The vapour of amyl alcohol which is passing through the jacket tube during the experiment may be generated by boiling the liquid in a glass flask, fitted with a cork and delivery tube.

The graduations upon the eudiometer tube used in this and the following experiment should not be made by scratching or etching, as this renders the glass liable to fracture at these points. Neither should they consist of rings of black paint or varnish (the favourite device of instrument-makers), as these are rapidly obliterated by the hot vapour or the steam. A good plan is to bind upon the tube, by means of fine copper wire, a band of copper foil, in the manner shown in fig. 35.

The bands may with advantage be fairly wide—say 8 or 10 millimetres. This not only renders them easily seen at a distance, but also allows a little margin for the errors which are incidental to such rough volumetric experiments.

FIG. 36.

75. This experiment is much more beautifully shown by means of the apparatus as modified by Hofmann. This consists of a straight eudio-meter tube about 1 metre long. Three gradua-tions, about 20 centimetres each, measuring from the closed end, are made upon this tube; just below the third graduation is fitted a cork which carries the jacket tube to surround the upper portion of the eudiometer (fig. 36). The eudio-meter is held in a clamp which works up and down an upright support by means of a rack and pinion, and it dips into a tall cylinder mer-cury trough, at the bottom of which a caout-chouc pad is cemented. The eudiometer is filled with mercury, covered with the thumb and inverted in the trough; being longer than the barometric column, the mer-cury will sink in the tube some distance. The tube is then sup-ported in the clamp, and screwed up until its mouth is only about one centimetre beneath the mercury. Dry electrolytic gas is then bubbled up *until the mercury stands about* 5 centimetres *above the lowest graduation.* The tube should then be lowered right to the bottom of the cylinder, and the excess of mercury allowed to overflow. As the gases in the tube are under reduced pres-sure, the temperature of 100°C. is sufficient to vapourise the water which is formed by the explosion, consequently the use of

amyl alcohol is obviated, and a current of steam is delivered through the jacket. When the steam is passing, the position of the tube is so adjusted that the gases occupy the space down to the third graduation. The height of the column of mercury in the tube above the level of that in the trough is indicated by a marker which also works up and down the upright support by a rack and pinion. The eudiometer is then lowered down to the pad, the marker remaining stationary, and the spark passed. The tube is then raised until the column of mercury is the same height above the level of that in the trough as it was before, which is indicated by the marker, and it will be seen that this is coincident with the second graduation.

FIG. 37.

To avoid the risk of the eudiometer being cracked by drops of hot water which come with the first steam that enters the jacket, it is well to arrange that the tube delivering the steam is so bent (fig. 37) that the steam shall be directed against one side of the jacket; and as a further safeguard the steam may be passed through a small empty flask immediately before entering the apparatus.

76. To show the volume composition of water analytically. The decomposition of water by means of an electric current may be shown by passing the current from 5 to 10 Grove's cells through water acidulated with sulphuric acid. A small cylinder bottle is fitted with a cork with three holes; into one is fitted a delivery tube, and into the others two short pieces of glass tube carrying the platinum electrodes (fig. 38). The electrodes may be made by welding strips of platinum foil on to platinum wires,[1] and fusing the wires into the ends of two pieces of glass tube. These tubes

FIG. 38.

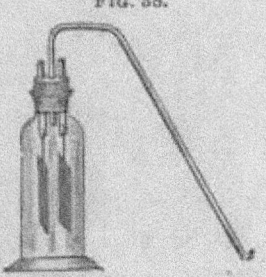

[1] For the method of welding platinum see Exp. No. 7.

are partially filled up with mercury, and contact with the battery is made by dipping the wire leads into the mercury.

77. This decomposition may be shown upon the screen, by means of a lantern microscope. The electrolysis is effected in a small cell capable of being held in the clamp of the microscope, and fitted with electrodes of fine platinum wire. Care must be taken to arrange both of the wires in the same plane, so that both may be in focus. If the wires are greased by passing them through the fingers moistened with oil, and the strength of current carefully regulated, the bubbles of gas which appear on the electrodes will grow to a great size before rolling up the

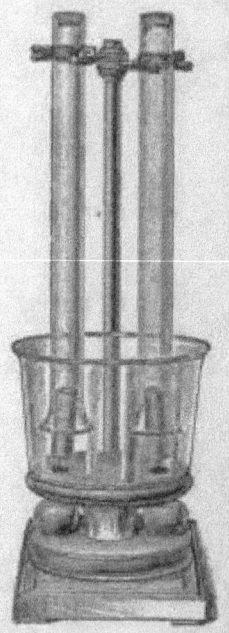

Fig. 39.

wire. To regulate the current a resistance cell must be placed in the circuit, consisting of a tube (conveniently a U tube) containing a solution of copper sulphate, and two copper wires which can be drawn nearer or farther from each other at will.

In the absence of a lantern microscope this experiment may be performed in a larger glass cell, using platinum plates as electrodes. The plates should be placed edgeways to the screen, and an image of them projected on to the screen by means of a short-focus lens. (See p. 325.)

78. To collect and measure the separate gases. For this purpose various forms of apparatus may be used. If it be desired to show the properties of the two gases, a convenient form of apparatus consists of a glass basin with two tubulures in the bottom (fig. 89) into which are fitted corks carrying the electrodes. The basin stands in a wooden support, and connection is made with the electrodes either by causing them to dip into small wooden cups containing mercury, which are fastened on the foot of the apparatus, or directly by

means of binding screws. Two cylinders of equal size are suspended over the electrodes.

79. A compact form of apparatus may be constructed in the following way. A wide-mouth glass cylinder (fig. 40) is loosely fitted with a cork with four holes. Through two holes pass glass tubes reaching to the bottom and recurved at the end; these carry the electrodes fused into their ends, and are filled up with mercury. Through the other two holes pass two short tubes which are blown upon the ends of wider tubes, capable of passing over the platinum electrodes; one of these is placed over each electrode. To the upper extremities of these two tubes delivery tubes may be attached, and the gas evolved from the two electrodes collected in a pair of graduated cylinders over a pneumatic trough, care being taken that each delivery tube dips to the same extent beneath the surface of the water in the trough.

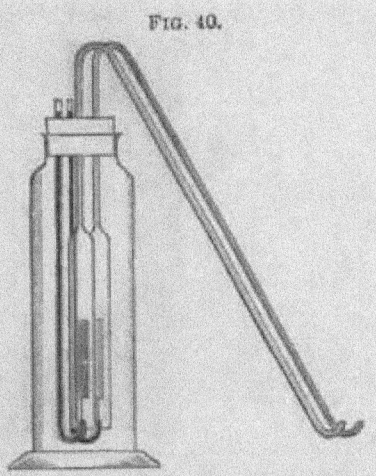

FIG. 40.

80. The relative volumes of the two gases as evolved by electrolysis may be shown upon the screen by projecting the image of a small electrolysis apparatus, constructed in the form of a U tube with a side tube blown in near the base (fig. 41), and with a short platinum wire fused into each limb of the U. The apparatus is made of such a size that the entire length of the tubes from the electrodes to the top is well covered by the lens employed to condense the light upon it; each limb may be graduated into two divisions by means of small rings of caoutchouc, cut from caoutchouc tube. Dilute acid is dropped in at the open tube by means of a drawn-out thistle tube, and

shaken up into the closed limbs by inverting the apparatus. The tubes being small, the acid will not run out, and when the current is passed the displaced liquid may be allowed to drop into a small dish. As the tubes may not be absolutely true in the

FIG. 41.

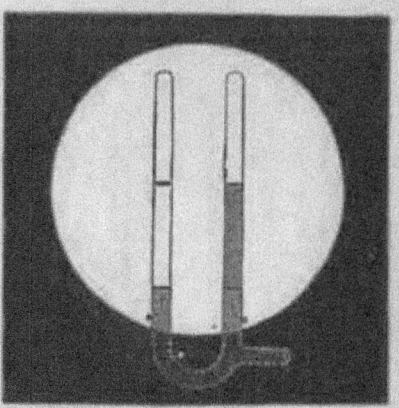

bore, it is well to find the exact places for the graduations by experiment, and then to mark, on the lower part of the tubes, the letters H and O, respectively.

81. The gravimetric composition of water. The principle of Dumas' method for determining the composition of water by weight, may be illustrated by passing a stream of hydrogen over copper oxide contained in a bulb tube, one end of which is drawn out and bent at an obtuse angle (fig. 42). This drawn-out end is fitted by means of a cork into one neck of a small two-necked receiver, or into a small Wurtz's flask, and the exit from the receiver, or flask, is attached to a chloride of

FIG. 42.

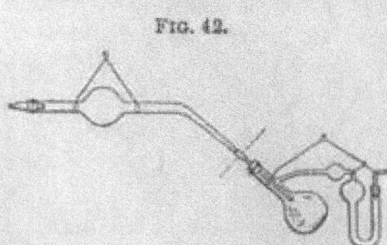

calcium tube. Before the hydrogen enters the bulb tube it is also passed through a drying tube. These drying tubes are merely to symbolise the apparatus for purifying and drying the gas. The hydrogen should be allowed slowly to pass through the apparatus, and when all air is expelled, the bulb containing the copper oxide may be gently warmed. The combustion of the oxide will be manifest by the glowing of the mass and its reduction to metallic copper. The bulk of the water will condense in the receiver.

If it be intended to show the change of weight undergone by the different parts of the apparatus, it is well to counterpoise the two essential portions, viz. (1) the bulb tube with its charge of copper oxide, and (2) the receiver with the attached calcium chloride tube, by means of small beakers containing either mercury or fine shot. In this way each part may be counterpoised beforehand, and, at the time of doing the experiment, may be shown to exactly balance its own tare. After the experiment has proceeded for a short time it is easy to demonstrate that the bulb tube has lost weight, and that the receiver has gained.

With a little care it is quite possible to make the experiment so far a quantitative one that, by noting the alteration of weight of the parts, the composition of water may be calculated. (See Exp. No. 378.)

82. To show the decomposition of steam by electric sparks. A small four-way globe has fitted into two opposite tubulures two corks carrying platinum wires fused into glass tubes. The third tubulure carries a delivery tube (fig. 43). By the fourth, the globe is connected to a flask in which steam is being generated. When the whole of the air from the apparatus has been expelled by the steam, a series of electric sparks is passed between the platinum wires within the globe, and at once it will be seen that the gas escaping from the delivery tube is no longer entirely condensed in the water of the pneumatic trough, but that small bubbles are collected. It is well to collect the gas in a long narrow tube instead of the usual cylinder or gas jar. The

lamp employed should be provided with a chimney, or in some way screened from any draught, as a momentary failure of the

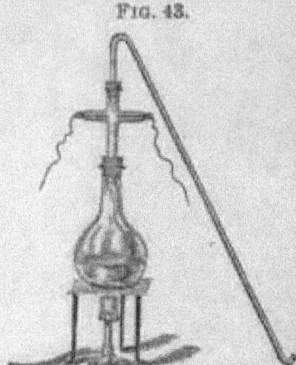

FIG. 43.

heat will cause the water in the trough to be driven back into the apparatus.

The four-way globe may be replaced by a four-way tube, which may readily be blown from glass tube; in this case the platinum wires can be fused at once into the two opposite openings. For this experiment a coil must be used capable of giving a strong spark of not less than one inch.

For decomposition of water by metals see Hydrogen.

For decomposition of water by chlorine see Oxygen.

83. To show the colour of water. A tube made of sheet tinned iron, 10 centimetres diameter and from 4 to 5 metres long, is provided with caps with plate glass ends; the caps may be permanently cemented on. The tube is also provided, near to one end, with a small opening, through which water can be introduced, and which is capable of being closed with a cork. The tube is placed in a horizontal position, and half filled with pure distilled water. A parallel beam of light is sent through the tube, and an image of the end projected upon the screen. It will be seen that the light emerging from the water is of a bluish-green colour. The disk upon the screen will therefore have its upper half green, the lower portion being the unaltered light from the lamp. After use, the tube should be carefully drained dry and then corked up to prevent the entrance of dust.

84. The same effect may be shown by causing a beam of light to travel twice through a shorter stratum of water. A glass cylinder about 10 centimetres diameter and one metre long (or longer if convenient) has a layer of clean mercury

placed in the bottom, and is then filled up with the purest distilled water. A narrow beam of light is reflected by a mirror
down through the water in such a way that the beam will be
reflected up from the surface of the mercury and may be received upon a second mirror and thrown upon the screen (fig. 44).

FIG. 44.

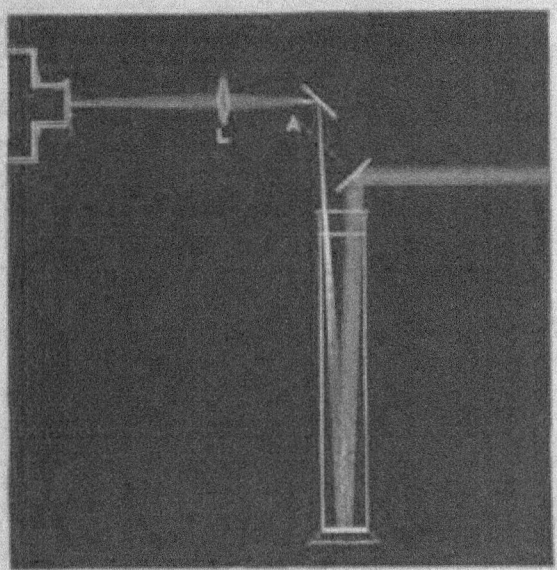

The greenish colour of the emerging light will be very per-
ceptible. This method has the obvious advantage that the
audience can see the colourless appearance of the water contained in the cylinder.

By intercepting the beam at A by a small piece of mirror held
in the hand, the light may be thrown direct on to the screen,
whereby the colour of the white light may be compared with
that of the beam, after its passage through the water.

In order that the light upon the screen may be defined, it
is well to employ a small stop upon the condenser in the
lantern, and to focus an image of it upon the screen by means

of a focussing lens (L, fig. 44). The beam from this lens is then diverted into the water.

85. **To show the spheroidal state of water.** A metal crucible, preferably nickel, has its bottom turned flat and polished in a lathe. If the crucible is not thick enough to allow of this, an iron disc may be brazed on the bottom, and the iron turned flat and smooth. Three or four holes are bored in the side of the crucible, so as to enable a flame to be kept alight under it when inverted. The crucible is placed upon a chimney carrier screwed to the tube of a Bunsen, and the whole is stood upon a levelling block, and placed in front of the lamp. A parallel beam from the condenser with a small stop is sent across the flat bottom of the crucible. A dropping funnel containing water, with its stem drawn out to a long capillary tube, is held by a clamp over the crucible, so that the point of the tube is about 2 millimetres above the metal, and an image of this tube is carefully focussed on the screen. A small jet of gas is lighted in the crucible, and water is allowed very slowly to drip from the funnel. The water will assume the spheroidal condition, and will be held from rolling away by the fine tube which delivers it. If the level of the crucible is rightly adjusted, a distinct line of light will be visible on the screen between the spheroid and the hot metal.

A simple and convenient form of levelling table is easily made in the following way :—

Three round-headed brass wood screws, about 3½ centimetres long, have soldered upon each of them, nearly up to the head,

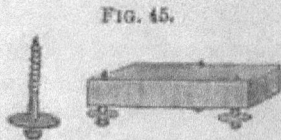

FIG. 45.

a brass 'blank' about 2 centimetres diameter (fig. 45). These are screwed into a mahogany block, 10 centimetres square by about 1½ centimetres thick, two of the screws being at the opposite ends of one side, and the third in the middle of the opposite side. The holes in the wood should be sufficiently large to allow of the screws being easily twisted with the fingers by means of the flange made by the brass blank.

86. The spheroidal condition of water may also be demonstrated by arranging the spheroid in an electric circuit, in which there is also an electric bell.

For this purpose a platinum dish or capsule is connected by a wire to one terminal of a small battery. A platinum wire held in a clamp is lowered down until its end is within 2 millimetres of the bottom of the dish; this wire is connected through an electric bell to the other terminal of the battery. The dish is made hot and two or three drops of acidulated water (sulphuric acid) are carefully introduced with a pipette. So long as the water remains in the spheroidal state, no contact with the dish results, and the circuit still remains open. On withdrawing the lamp and allowing the dish to cool, the spheroid breaks down and wets the dish, so closing the circuit and causing the bell to ring. Care must be taken that the wire is so near to the dish, that when the drop of water (which will be considerably diminished by evaporation) flattens itself down upon the dish, as the latter cools, the wire will still be dipping into the liquid.

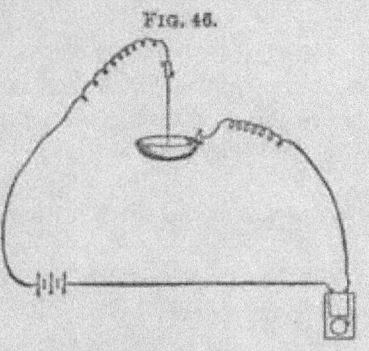

FIG. 46.

87. To show the maximum density of water.[1] A piece of thin glass tube is closed at one end like a test tube, and drawn out to a small neck at the other. The tube should be capable of holding about 100 cubic centimetres. Into the neck is fitted a straight piece of thermometer tube, the neck being made of such a size that the thermometer tube with a small piece of caoutchouc tube over it can be firmly squeezed in. In this way

[1] See Table III. in the Appendix.

will be obtained what may be regarded as a thermometer with a very large bulb. In order to compensate as far as possible for the expansion and contraction suffered by the glass in the course of the experiment, a quantity of mercury is introduced into the bulb, the volume required being calculated from the co-efficients of expansion of glass and mercury. For a vessel of a capacity equal to 100 cubic centimetres, 13 cubic centimetres of mercury will be required. The bulb is then completely filled with water and the stem inserted. The whole is then placed in a vessel of water, at about 7° or 8°C., until its temperature has reached that point. The apparatus is placed in front of the lamp, so that the water in the stem is within the field, and an image of the stem thrown upon the screen, alongside of a vertical scale. Should the water in the stem not be standing at a convenient height, it can be brought to any desired point by screwing the stem slightly in or out of the neck of the bulb. The vessel of water is then replaced by one containing finely broken ice. The contraction which follows will be indicated by a receding movement of the water in the stem ; in a few minutes this move-ment will stop, and the liquid will be seen to again expand.

The experiment may be performed in the reverse way. The water may be cooled to 0°C. by the bulb being immersed in ice for some time, and the ice replaced by water at about 8°C., when the first effect of the warmer water will be to cause a contraction in the volume of the water in the bulb. The water used should be only a very few degrees above the point of maximum density, or the expansion of the glass will make itself felt before that of the mercury, and the compensating effect of the latter will be lost.

88. This may also be demonstrated by means of a float. A small float is blown out of thin glass tube, having for its stem an extremely fine rod of glass, about the thickness of an ordinary pin. The float is so weighted with a drop of mercury that when placed in water at 0°C. it floats five or six centimetres below the surface, and with about the same length of stem

projecting above the water. The weight of the apparatus may be finally adjusted by either snipping a piece of the stem off, or by fusing a little piece on. The float is placed in water contained in a test tube which is supported in front of the lamp, and an image of the point of the stem projected upon a vertical scale placed against the screen. The test tube is then surrounded with crushed ice contained in a beaker, until the temperature has fallen to 0°C. On removing the ice and carefully replacing it by a beaker of water at about 8°C. without disturbing the float, the latter will be seen slowly to rise until a maximum is reached, and then to fall again. Great care must be taken that the float does not touch against the sides of the vessel, or its movement will be altogether stopped. The experiment is a little critical.

89. **To show the high specific heat of water; comparison with mercury.** This may be done in a calorimeter, and the effect shown upon the screen. A convenient apparatus may be arranged by cutting the narrow limb from a small Bunsen's calorimeter, and replacing it by a piece of flat-bore thermometer tube, which may be connected to the bulb by a piece of caoutchouc tube 5 or 6 centimetres long. The bulb is filled with water, which may be coloured slightly with aniline blue to render it more visible in the stem, an image of which is thrown upon the screen. The height of the water in the stem can be brought to any convenient point by screwing the rubber connection farther on or off the glass tube. Equal weights (10 grams is convenient) of mercury and water are placed in test tubes and heated in a beaker of boiling water. On pouring the hot mercury into the calorimeter, only a small expansion will be noticed, but on adding the water, the expansion will cause the image of the liquid to go right across the screen. The mercury should be introduced *before* the water, and it may either be left in while the water is added, or it can be tipped out without disturbing the stem, owing to the flexible connection.

90. To show the expansion of water on freezing. This may be done by means of a similar apparatus to that described for showing the maximum density. The bulb, which for this purpose may be much smaller, is immersed in ice-cold water, and an image of the stem and the liquid projected on the screen. The ice water is then replaced by a freezing mixture, and almost immediately the expansion will be manifest.

91. This experiment is perhaps more strikingly shown by using a small Bunsen's ice calorimeter, with a moderately fine stem, and of such a size that the bulb and stem will fall within the field of the lantern lens. The apparatus is completely filled with water, which must stand at such a height in the stem that its movements will be well within the field. About 5 cubic centimetres of ether are placed in the inner tube of the calorimeter, and a few fragments of solid carbon dioxide added. Ice rapidly forms, as a thimble, round the closed end of the inner tube, and each addition of a little piece of the solid carbon dioxide is accompanied by an immediate further expansion. If a small sponge attached to a wire is pushed into the tube, the ether may be quickly mopped out, and on pouring in a small quantity of warm water, the ice will be seen to melt, and the consequent contraction will be evident.

It is not necessary to cool the water in the apparatus before the experiment, and doing so is liable to cause moisture to deposit on the outside, and so obscure the image.

92. By causing the expansion to burst the vessel containing the water. A small iron bottle (sold by instrument makers for the purpose) is completely filled with water, which has been recently boiled, and then cooled. The plug is then firmly screwed home. The bottle is then buried in a freezing mixture contained in a wooden bowl, and covered over with a cloth or wooden cover. After a short interval the bottle will burst, with a dull thud, and may then be withdrawn.

Occasionally these bottles have been known to burst with great violence, and fly with considerable force to distant parts of the room, hence the desirability of covering the bowl. This explosive bursting only takes place, however, when the bottle has not been completely filled with water; if care be taken to avoid inclosing any air when the plug is screwed in, the bursting of the vessel will not be attended with any violence.

93. The longitudinal fracture which usually results from this bursting, may be illustrated by bursting a piece of 'compo' pipe. A piece of such pipe has one end closed by flattening it with a hammer and turning the flattened part back upon itself. The other end is nearly closed up, the pipe filled with water, and the opening hammered together. When such a pipe so filled with water is placed in the freezing mixture it will be split along the side.

A glass tube filled with water and hermetically sealed will also show this effect well. The glass tube should be simply laid upon the freezing mixture, so that it can be removed afterwards without being disturbed.

94. To show the supercooling of water and rise of temperature on solidification. The apparatus required consists of a thermometer, the bulb of which is inclosed in a larger glass bulb, nearly filled with water, and rendered vacuous. To the stem of the thermometer is attached a flat glass graduated scale, so that an image of the scale and the thread of mercury may be thrown upon the screen.

The apparatus may be made in the following way.[1] A thermometer, known as a 'paper scale thermometer,' is taken. The special point about this particular thermometer is that its real stem is an extremely small glass tube, which is inclosed in a wider glass tube, in which a paper scale is fixed. The outer tube is cut off at a point about 8 centimetres above the bulb by scratching with a file at that spot, and touching with a red-hot point of glass, and the scale removed (fig. 47). A piece of wide glass tube closed at one end is drawn out at the other to a slightly conical neck, the opening being of such

[1] The complete apparatus may be obtained from F. Müller, Bonn.

a size that the outside tube of the thermometer stem with a piece of thin caoutchouc tube stretched over it may be tightly squeezed into it. Upon the shoulder of this tube, near the neck, a small side opening is made (fig. 47) by directing a fine blow-pipe flame upon it, and drawing out the glass by means of a piece of platinum wire. (See Exp. No. 60.)

This tube, which may be about 10 centimetres long, is about three parts filled with distilled water, and the thermometer pushed into it until the bulb is about 1 centimetre from the bottom. The water is then carefully boiled, and as the steam issues from the small side orifice the point is sealed by means of a fine blow-pipe flame. A flat strip of thin glass about 4 centimetres long is laid upon the paper scale which was taken from the thermometer, and the graduations traced upon the glass with a writing diamond (or they may be etched with hydrofluoric acid).

FIG. 47.

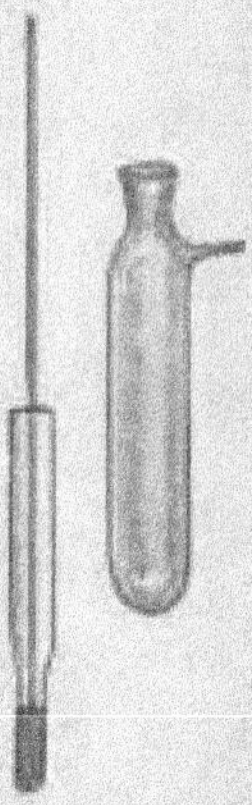

The apparatus is placed in broken ice for about half an hour, and while there the scale may be attached to the fine stem by means of a small touch of cement, the zero of the scale being made coincident with the point at which the mercury is standing. The apparatus is now ready for use. It is held by a clamp in front of the lamp, and an image of the stem and scale thrown upon the screen. The bulb is then placed in broken ice until the thermometer registers 0°C. The ice is then removed, and a freezing mixture at about −10° substituted until the thermometer has fallen to about −8° or −9°.

By the application of freezing mixtures of successive degrees of cold, the temperature of the water may be lowered −12° or even −15°C. without solidification taking place, the important point being to apply the cold in stages, as the sudden application of a very cold mixture invariably determines the solidification of the water. The lower the temperature to which the water is reduced the more critical it becomes, so that a slight shock will often cause it to freeze; but at about −8° to −10°

the water may be shaken even fairly roughly without its
becoming solid. In performing the experiment it is best not
to take the temperature below about −9°, and, instead of
endeavouring to solidify it by shaking it (which is not only
uncertain, but disturbs the image on the screen so much that
any movement of the mercury in the thermometer is rendered
invisible), it is best to apply a freezing mixture at about
−20° or lower, when ice instantly forms, and the mercury rises
to the zero.

The freezing mixture most commonly used is ice and
common salt, which, without any special care in the preparation, will give a temperature of −18° to −20°C. By far the
most convenient mixture for lecture purposes, where as little
loss of time as possible is a desideratum, is made by adding to
crushed ice about an equal volume of crushed crystallised
chloride of calcium. This mixture when roughly made will
give a cold of −28° to −30°C. The materials on mixing immediately become liquid ; the mixture, therefore, is more easily
manipulated, and tubes and objects immersed in it, coming
more uniformly into contact with it, are much more quickly
cooled than in the nearly solid mixture of ice and salt. To
bring the temperature of the mixture up to any desired point
small quantities of water may be added until a thermometer
registers the desired temperature.

95. To show that water boils at lower temperatures when
under reduced pressure. A large round-bottomed flask about 1½
litres capacity is tightly fitted with a caoutchouc cork carrying
a stopcock. About half a litre of distilled water along with
two or three scraps of platinum are put in the flask and heat
applied. When the water boils, the lamp may be removed, and
the flask attached to a small exhaust pump. Each stroke of
the pump will be followed by renewed vigorous ebullition of the
water.

96. The reduction of pressure may also be effected by cooling the flask. The water is first boiled briskly to expel the air

from the vessel; the lamp is then withdrawn and the cock closed. On applying a wet sponge to the upper part of the flask, condensation of the inclosed steam takes place, resulting in a partial vacuum, and the water is seen to boil vigorously at each application of the cold sponge.

97. This same phenomenon is well seen in the following experiment, viz., the freezing water by its own evaporation in the 'Carré' freezing machine. The apparatus consists essentially of an air-pump and desiccating chamber, and is so arranged that the air as it is withdrawn from the vessel to be exhausted is made to pass over a layer of sulphuric acid contained in the desiccator.

As full and minute directions are always supplied with the machine, only a few hints will be necessary. The water to be frozen is placed in a decanter-shaped bottle, which should be filled only to a depth of about 2 centimetres. While exhausting, it is well to disconnect the lever which moves the stirrer until the gauge shows a pressure of about 2 millimetres; the stirrer may then be gently worked by hand, avoiding splashing, when the water will enter into violent ebullition, and in a few moments, and without further pumping, will solidify. The sulphuric acid should be put in fresh each time the machine is used, or, at most, not allowed to remain in longer than a few days.

The pump must be in such a working condition that it is capable of reducing the pressure to 1 millimetre. After charging the apparatus with acid, it is therefore well to close the cock communicating with the decanter, and ascertain if this degree of exhaustion is obtainable; if not, it is useless to attempt the experiment of freezing the water, although it may be made to boil.

98. To freeze water by its own evaporation, using an ordinary air-pump. This experiment may be beautifully shown in the following way, by means of any air-pump capable of exhausting to within 2 or 3 millimetres.

For this purpose a tall bell-jar (about 60 centimetres high), with a wide mouth at the top, is fitted with a caoutchouc stopper with three holes. Into one hole is fitted a stoppered dropping funnel, the end of which is cut off (about 5 centimetres below the caoutchouc) at an acute angle, and to which is tied a small piece of lamp cotton. Into the second hole is passed a stout glass rod, or a piece of tube closed at the ends, passing to within 2 or 3 centimetres of the bottom of the bell-jar; the hole for

this tube need not pass more than about half way through the caoutchouc, the tube being pushed in from the under side. Round this tube is wound, as a spiral, a thick thread of asbestos, obtained by unravelling a piece of asbestos cloth. The thread may be secured at the top and bottom of the tube by means of a bind of thin platinum wire. Into the third hole a similar funnel tube to the first is fitted, but with its end drawn out to a fine tube and bent at an obtuse angle so that the point just touches the long tube.

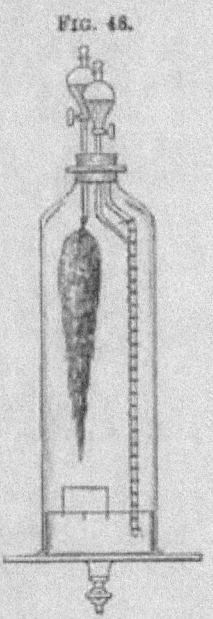

FIG. 48.

The bell-jar with its fittings is placed upon the plate of an air-pump, over a shallow glass basin nearly as wide as the bell-jar. A smaller glass basin, standing on a wire triangle over the larger one, is placed immediately under the suspended cotton wick. The funnel with the drawn-out end is filled with concentrated sulphuric acid; the other funnel being filled

with water. The bell-jar is then exhausted, and when the pressure has reached about 3 millimetres, two or three drops of water are allowed to enter from the funnel, so as to wet the cotton wick; any water which falls will be received in the small basin. A gentle stream of sulphuric acid is next allowed to enter, and it will flow round and round the tube, upon the spiral of asbestos, thereby presenting a large surface of acid. The drop of water upon the wick will almost immediately freeze; water is then cautiously admitted at such a rate that each drop, as it runs over the already formed ice, congeals.

carefully regulating the supply of acid and the rate of admission
of water, the whole of the latter may be frozen as it enters,
forming a beautiful icicle, which can be obtained 20 or 30
centimetres long and several centimetres thick.   The longer
the icicle becomes the more readily does the incoming water
freeze as it runs over it.   At the beginning of the experiment a
little care is necessary to so regulate the freezing (by controlling
the flow of both the acid and the water) that the ice does not
form in the stem of the funnel, and so stop the ingress of more
water.   Should this happen, the flow of acid should be stopped,
and a beam of the electric light concentrated on the ice for a
few moments, or a vessel of hot water may be held near to the
glass.

99. By a slight modification of the experiment the air-pump
may be entirely dispensed with, and a vacuum obtained without
its use. In this case the tube upon which the asbestos is wound
is open at the bottom, passing through the cork, and ending
above in a stopcock. A Wurtz flask, of about 200 c.c. capacity,
is fitted with a caoutchouc stopper carrying a small dropping
funnel, whose stem reaches nearly to the bottom. The branch
tube of the Wurtz is bent over so that it may be attached to the
stopcock upon the long tube. This joint must be wired. The
bell-jar is placed upon a stout piece of plate-glass, instead of
the air-pump plate, and a stream of ammonia gas passed through
the entire apparatus. The ammonia is admitted through the
funnel which is destined to supply the water (B, fig. 49), and
it passes out by means of the long tube reaching to the bottom
of the receiver, through the Wurtz flask, and may be conducted
into a vessel of water. Pieces of glass tube, drawn to a taper,
with short pieces of caoutchouc slipped over them, may be
fitted into the tubulure of the funnels, in order to pass the
ammonia in and out. When the air is all expelled, which is
ascertained by noting how the escaping gas is absorbed in the
water, the stopcock of funnel C is opened for a few seconds, in
order to sweep out any air remaining in the stem of this funnel,

and again closed. The apparatus being now completely filled with ammonia the stopcocks are all closed, and the ammonia generator disconnected. Funnel B is filled with water, and C with concentrated sulphuric acid.

The next step in the operation is to absorb the ammonia, which is effected by means of sulphuric acid, and the absorption is performed in the Wurtz flask. 2 or 3 cubic centimetres of acid, previously diluted with about its own volume of water, are allowed to enter by the small funnel D. It is necessary to dilute the first portion of acid in this way, or the stopcock would be instantly blocked up with sulphate of ammonia; but, after the first few drops have passed in, the funnel may be filled up with strong acid. The ammonia in the flask is almost instantly absorbed, and on opening the cock A, the gas within the bell-jar is rapidly drawn out, and in a short time, by the gradual admission of the acid into the Wurtz, the whole of the ammonia is absorbed. A small mercury gauge, previously placed in the bell-jar, will enable the experimenter to see the extent to which the pressure is being reduced. When a sufficiently good

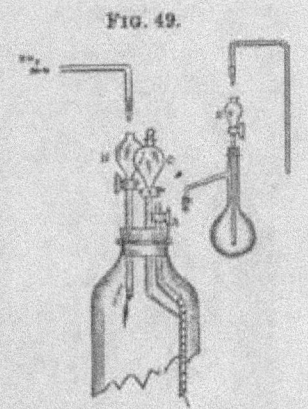

FIG. 49.

vacuum is obtained the stopcock A may be closed, and the Wurtz flask removed. Any minute traces of ammonia still present will be instantly absorbed on the first introduction of acid from funnel C in the process of the experiment. If more than traces of ammonia are left in the bell-jar when the acid from C is passed in, fumes of ammonium sulphate will be formed, which will deposit on the glass and obscure the experiment, and so defeat the whole object of conducting the absorption outside the bell-jar.

The experiment is then proceeded with exactly as described in No. 98.

100. The freezing of water by its own evaporation may also be shown in an ordinary cryophorus. The apparatus consists of a long glass tube with a bulb at each end; it is partially filled with water, completely exhausted of air and hermetically sealed. The tube is so bent that one bulb can be conveniently immersed in a freezing mixture, while the water is mostly contained in the other. In a few minutes the water will be seen to become cold by the deposition of dew on the outside, and in a short time ice will form on the surface of the water. If the water that distils into the cooled bulb does not at once freeze, the apparatus should be gently shaken, but afterwards it should be allowed to remain undisturbed.

101. To illustrate the fact of cold being developed by evaporation. The bulb of an air thermometer is tied up in a thin mus-

FIG. 50.

lin bag and immersed in water, and the position of the liquid in the stem noted. The bulb is removed and a stream of air from a small foot-bellows projected against it, when the effect of cold will be seen by the movement of the liquid in the stem. An air thermometer is conveniently made by connecting a thin glass bulb to one limb of a tall U tube containing coloured water, by means of a piece of caoutchouc tube about 10 centimetres long. A side tube (fig. 50) should be attached to the apparatus, which can be closed by a pinch-cock, so that the levels of the liquid in the two limbs can be adjusted at any temperature.

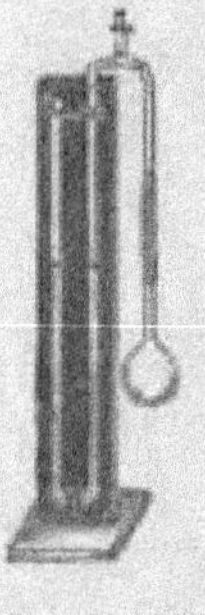

102. A small quantity of a volatile liquid (benzene, carbon disulphide, ether, &c.) is placed in a little flask and corked. This is stood on the face of a thermopile which is attached to a galvanometer, and the position of the needle noted; on pouring a drop or two of the liquid from the flask upon the pile, an instant deflection in the direction of cold takes place.

103. About 20 cubic centimetres of ether are placed in a small flask, which is stood upon a wooden block upon which a few drops of water have been poured. A stream of wind from a foot-bellows is blown through the ether and in a few moments the water will be frozen, and the flask will be firmly cemented to the block. (See also Sulphur Dioxide, Exp. 599.)

104. To show convection currents in water. This may be shown upon the screen by pouring into some water, contained in a narrow cell in front of the lamp, a small quantity of a mixture of carbon disulphide and paraffin oil. This mixture is made to have the same density as water, so that on stirring with a glass rod the small oily globules will remain suspended throughout the body of the water. Heat is applied near to the bottom of the cell by sending a current from about ten Grove's cells through a short platinum wire which connects two stout copper wires, and the currents set up in the water will be rendered evident by the movement of the oil globules.

Fig. 51.

15 cubic centimetres carbon disulphide with 17 cubic centimetres paraffin will be found to be very nearly what is required to yield a mixture with the same density as water, with average samples of paraffin oil.

Sometimes sawdust, or other light solid substances are employed, but not with so good a result as the liquid globules; still less advantageous is it to indicate the current by means of soluble colours, as in this case the whole mass of water is rapidly reduced to a uniform tint.

105. To show that water is a bad conductor of heat. A fragment of ice is weighted by twisting a small piece of lead wire round it, and placed at the bottom of a test tube nearly filled with ice-cold water. The tube is then held over the flame of a Bunsen so that the water at the top is being heated; in this way the surface water may be boiled without the ice becoming visibly diminished. The water may be quickly emptied out and the piece of ice thrown on to a plate.

106. To show the formation of 'ice flowers.' A slab of ice (about $2\frac{1}{2}$ c.m. thick) as clear and free from bubbles as possible, is sawn from a block. The slab must be cut parallel to the plane of freezing, which is ascertained by the direction in which the streaks of air bubbles lie in the block. The slab is supported in a converging beam from the electric lamp, and an image of it projected upon the screen. In a few moments, as the warmth from the lamp is felt by the ice, a number of star-shaped 'flowers' make their appearance which will gradually increase in size as the ice continues to melt.

107. To show the effect of pressure upon the freezing-point of water. A large block of ice is supported on a stand and a fine steel wire slung over it. To the wire are attached heavy weights (about 10 kilos, or 21 lbs. will be found convenient), and it will be seen that the wire will slowly cut its way through the block, the ice melting in front of the wire owing to the pressure, and freezing again behind it as the pressure is released.

108. Two smooth slabs of ice may be pressed together by the hands, when they will be seen to adhere firmly to each other.

109. A striking way of showing this is to float three or four small fragments of ice in water contained in a flat-bottomed glass cell, and to project an image of them upon the screen, using the horizontal projector (see Lantern Illustrations). By means of a thin wire the floating pieces of ice may be made to bump gently against one another, when they will be seen to adhere at the points of contact. A string of them may be so joined together and drawn round the cell by the wire.

The same effect may be shown by floating larger masses of ice in a large trough or tank of water.

110. The solvent action of water upon solids.[1] To show different degrees of solubility, using as an example potassium chlorate and magnesium sulphate, 100 cubic centimetres of water are placed in two small flasks; to one is added 5 grams

[1] See Table IV. in the Appendix.

of finely powdered potassium chlorate, and to the other 90 grams of magnesium sulphate similarly powdered. The flasks are corked, and the contents agitated. In a few minutes the whole of the salt in each case will be dissolved. A small quantity more of each salt may be added to its respective flask to show that the water is saturated.

111. **To show increased solubility with rise of temperature.** The flask containing the chlorate (Exp. 110) is warmed, and a further quantity of the salt added in small quantities, until a hot saturated solution is obtained : on cooling, or on pouring a small quantity of the solution into a cold beaker, the deposition of the excess of salt may be shown.

112. **To show the formation of crystals upon the screen.** The salts which may be used with advantage are ammonium chloride, and potassium ferricyanide. Cold saturated solutions are made, which are best used unfiltered, as minute particles of foreign matter form nuclei for the crystals to start from. A quantity of the solution is poured over one side of a clean glass plate, which is then placed in the microscope. After a few moments when evaporation has proceeded a little, crystals begin to deposit, and grow in beautiful forms across the screen. It is essential that the glass plates used should be perfectly clean, or the solution will run off the surface in streaks ; the best method for cleaning them is to rub the surface with a wet cloth, and a sprinkle of finest washed emery powder, rinse them in distilled water, and place them in a dish of water until required. They are then taken out, one side only wiped dry with a clean cloth, and the solution of the salt poured upon the *wet side*, the excess of the solution being allowed to run off without getting upon the dry side. (See p. 327.)

113. **To show the production of cold on solution of a salt in water.** For this purpose ammonium sulphocyanate is a convenient salt. 20 grams of the salt are thrown into a small flask, placed upon a block of wood upon which a drop or two of water

F

has been poured; 25 cubic centimetres of water are poured upon the salt, and the contents gently shaken round, the flask being still upon the wetted block. In a few moments it will have become frozen to the block. The temperature produced by the mixture of this salt with water in these proportions is $-13°C$.

114. To show supersaturation. 500 grams of sodium sulphate are placed in a flask, and 200 cubic centimetres of water added. The flask is placed in a water-bath, and gently warmed until the salt is entirely dissolved. The mouth of the flask may either be tied over with parchment paper, or a plug of cotton wool may be placed in the neck, and a piece of paper folded over the mouth. When perfectly cold a crystal of the salt is dropped into the solution, and the mass will solidify.

The rise of temperature which accompanies the crystallisation may be demonstrated by placing the face of a thermopile against the flask; or it may be shown by dipping into the flask a tube containing a mixture of triethylamine and water, which will be rendered milky. (See Exp. 129.)

The phenomenon of supersaturation may also be illustrated by using sodium thiosulphate (sodium hyposulphite). A quantity of the salt is placed in a flask which is heated in a water-bath. The salt melts in its own water of crystallisation, and no addition of water is necessary.

115. To show decrease of solubility by increase of temperature. A convenient salt for this purpose is calcium citrate. A strong solution of ammonium citrate is made by dissolving 20 grams of citric acid in 25 cubic centimetres of water, and adding strong ammonia solution ('880) until the mixture is neutral to test paper (approximately 15 cubic centimetres of the ammonia will be required). A solution of calcium chloride is made by dissolving 40 grams of the crystallised salt in 40 cubic centimetres of water. Equal volumes of these two solutions are mixed in the cold. The precipitate which appears at first on adding one solution to the other rapidly dissolves on slightly shaking the test tube. On dipping the test tube into boiling water for an instant, a precipitate of calcium citrate forms in contact with the warm glass; on shaking, this again disappears. This phenomenon may be repeated two or three times. On leaving the tube in the hot water for a few moments the contents become perfectly solid, so that the tube may be inverted. Solutions of the above degree of concentration when mixed will gradually deposit the calcium citrate even in the cold, so that they must only be mixed when required.

116. To show the loss of water of crystallisation at ordinary temperatures. A number of crystals of sodium carbonate or sodium sulphate may be placed upon a clock glass standing over a dish of strong sulphuric acid, and the whole placed under the receiver of an air-pump, and the air exhausted. In the course of a few minutes the crystals will be seen to be rapidly efflorescing, and becoming opaque white.

117. To show the abstraction of water of crystallisation by sulphuric acid. A crystal of copper sulphate is suspended by means of platinum wire in a beaker containing strong sulphuric acid. The crystal almost immediately loses its blue colour, being covered over with nearly white anhydrous salt. If a large crystal be employed, and only partially immersed in the acid the contrast is more strikingly seen.

118. To show the loss of water of crystallisation by heat. Copper sulphate heated in a platinum dish loses water, and is converted into a nearly white powder.

119. A dilute solution of cobalt chloride, brushed upon paper, when dried and warmed, turns blue.

120. A solution of cobalt iodide, obtained by adding potassium iodide to cobalt chloride, when dried upon paper becomes of a brilliant green. These solutions may be employed to colour designs, the shades of pink given by the chloride and iodide being changed to blue and green when the design is exposed to heat.

121. A solution of magnesium platino-cyanide (a scarlet salt which dissolves in water to a colourless solution) evaporated to dryness in a porcelain dish deposits the red salt, containing seven molecules of water. If the dish be gently warmed, the salt loses a molecule of water, and becomes of a brilliant yellow colour. On still further heating, the salt loses four more molecules of water, and becomes white. To make this experiment effective, the dish must be carefully cleaned with fine washed emery (see Experiment 112) and three or four cubic centimetres of the solution of the salt poured in while the dish is still wet. The dish is then warmed over a flame, and the solution run over the surface constantly, so as to obtain a thin film of the salt coating the entire dish (see Experiment 124).

122. The re-hydration of salts, from which the water has been removed. Dry copper sulphate may be sprinkled upon a piece of moistened filter paper placed upon a white glazed tile, or on a white plate. The white powder is instantly turned blue.

123. A solution of cobalt chloride in alcohol, which is of a deep bluish-green colour, when mixed with water becomes pink.
An alcoholic solution, to which just enough water has been added to turn it pink, when warmed, will again become blue.

124. A dish prepared as described in Experiment 121, if carefully covered with a glass bell-jar, may be allowed to go cold, and the salt preserved in the white condition for some little time. If the dish now be gently breathed into, the entire mass will be instantly turned red.

125. The solvent action of water upon liquids. A quantity of water and ether may be shaken together in a stoppered bottle, and the two liquids separated by means of a separating funnel, the lower aqueous liquid being received into one bottle, and the upper and ethereal layer in a second. To show that the water has dissolved some of the ether, a quantity of the

aqueous solution is placed in a small flask fitted with a cork carrying a straight piece of glass tube. On gently warming the solution, ether vapour will be evolved, and may be ignited as it issues from the tube.

126. A striking way of showing the presence of the dissolved ether is to place some of the solution in a test tube, and immerse it in a freezing mixture until it is entirely frozen. The solid mass is removed from the test tube by dipping the tube for an instant in warm water. If this rod of ice be stood up on end upon a block, and a lighted taper applied to the top of it, it will inflame and continue burning until the ice has entirely melted.

127. The ethereal solution obtained above may be shown to have dissolved water by pouring a small quantity of it into a dish containing magnesium platino-cyanide, which has been heated until it has become white (see Experiment 121). The white salt will immediately re-hydrate itself at the expense of the water present in the ether, and become converted into the scarlet salt.

128. To show increased solubility of a liquid in water by rise of temperature. Phenol (carbolic acid) is a convenient liquid to employ.

Two or three cubic centimetres are poured into a test tube of water, and the mixture shaken together, so as to produce a milky solution. On dipping the test tube in warm water the solution will become clear, and, on again cooling, the milkiness at once reappears.

129. To show decrease of solubility of a liquid in water by rise of temperature. Equal volumes of triethylamine and water are mixed together, and the mixture slightly cooled; in a few moments complete solution takes place. On immersing the tube, containing this solution, into warm water, an instant turbidity is produced, owing to the diminished solubility. The mixture in the proportions given is extremely sensitive; at

20°C. it begins to show a distinct milkiness. The warmth of the hand is sufficient to cause a copious turbidity.

A quantity of this liquid sealed up in a glass tube constitutes a very convenient means for showing a slight rise of temperature (see Experiment 114). Mixtures of water and triethylamine in different proportions may be made, which require different degrees of temperature to cause the turbidity to appear.

130. To show the effect of pressure upon the solubility of liquids in water. A small quantity of the solution of triethylamine in water, as made above, is introduced into one of the compression tubes of a Cailletet apparatus (see Liquefaction of Gases). The tube may be supported by a clamp, as shown

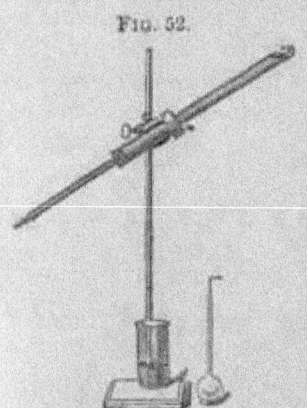

Fig. 52.

in the figure, and nearly filled with mercury by means of a thistle funnel drawn out and bent at the end. Four or five cubic centimetres of the solution are introduced in the same way, to completely fill the tube, leaving as little air as possible. By covering the end of the tube with the thumb, and inverting it, and gently bumping the hand down upon the table, the whole of the solution will be obtained in the capillary portion of the tube. The tube is then secured in the steel bottle of the Cailletet apparatus, and is surrounded with water at such a temperature as to cause the turbidity just to begin to appear. On then subjecting the contents of the tube to a pressure of from 20 to 50 atmospheres, the liquid becomes clear, and on releasing the pressure it once more becomes turbid. An image of the tube may be projected upon the screen.

NOTE.—The effect of pressure upon the solubility of one liquid in another may be illustrated by various mixtures.

e.g. glacial acetic acid has added to it a small quantity of carbon disulphide, drop by drop, forming a clear solution. On the addition of a few drops of water the solution becomes turbid; by adjusting the solutions a mixture may be obtained, which is clear at the ordinary temperature, but which on cooling to a particular degree will become turbid. If this solution be exposed to pressure, when the temperature is maintained just above the critical point—*i.e.* the point at which it begins to become turbid—a pressure of about 50 atmospheres will cause a turbidity. In this case the effect of pressure is to decrease the solubility.

If absolute alcohol be mixed with half its volume of water, and chloroform added, the latter liquid is at first entirely dissolved : when nearly as much chloroform has been added as there is water present, the liquid becomes turbid. On warming the mixture it clarifies, and on allowing it to cool it finally separates into two layers. If the warm solution be introduced into the compression tube, and there allowed to cool and separate into two layers, it will be found that the two liquids are affected by pressure in opposite directions.

The tube is cooled until the bottom layer just begins to show turbidity, the upper layer remaining clear, and a pressure of about 150 atmospheres applied. The lower turbid liquid becomes clear, and the upper clear solution becomes turbid; on releasing the pressure the solutions return to their former condition—that is, the upper liquid clarifies, and the lower one again becomes turbid.

131. The solvent action of water upon gases.[1] A glass globe with a long neck may be filled with either ammonia or hydrochloric acid gas by displacement ; during the filling, the escaping gas should be passed into water, in order that the degree of purity of the gas may be seen by the completeness of the absorption that takes place.

When the globe is full it should be securely corked up,

[1] See Table VII. in the Appendix.

and if the corked end be stood in a vessel containing a little mercury it may be kept any length of time until required. The cork is withdrawn under water in a large trough or basin (the water may be coloured with litmus), and the neck of the flask firmly held, for as soon as the water enters the globe it rushes up with a considerable shock. Unless the globe is fairly strong it should not be completely filled with the gas, or there is some risk of the uprush of water fracturing it.

A short-necked globe may be employed, fitted with a good india-rubber cork carrying a narrow glass tube reaching to within about six centimetres of the top of the globe. The water will in this way be forced up as a jet or fountain, and the process of absorption will not be so violent.

In order to start the absorption by getting the water to rise in the tube, it may be necessary to cool the globe by pouring upon it a few drops of ether.

132. This experiment may be shown on a smaller scale by filling a cylinder with either of the gases named, and opening it under water. The mouth of the cylinder must be covered with a thick glass plate, made of plate glass, for if a thin plate be used the moment it is shifted so as to admit the first drop of water the force of the inrushing water will certainly fracture the plate and the glass is liable to cut the hand of the operator. A stoppered vessel must not be used.

133. To show the solubility of a less soluble gas, such as chlorine or sulphuretted hydrogen. A stoppered cylinder is filled with the gas, and a small quantity of water introduced; this is well shaken up with the gas, and the cylinder opened under water. A quantity of water will be seen to enter. This is again agitated with the gas, and the cylinder again opened. In three or four operations the whole of the gas will be absorbed.

134. To show the effect of pressure upon the solubility of a gas in water. Two glass tubes about 1 metre long and 18 millimetres bore, closed at one end, are connected by means of

a T piece and flexible tube to a mercury reservoir (fig. 53). Care must be taken to select glass tubes of fairly uniform bore throughout. One of the tubes is filled with air, the other contains a short column (about 50 millimetres) of water which has been saturated with ammonia, and the tube is filled with ammonia gas. The cork carrying the air tube is fitted with a stopcock, so that air may be introduced or withdrawn as may be necessary to adjust the levels of the liquid in the two tubes. The level of the mercury in the air tube should be made to coincide with the upper surface of the water in the other tube when the liquids are down at the lowest division marked upon the scale behind the tubes.

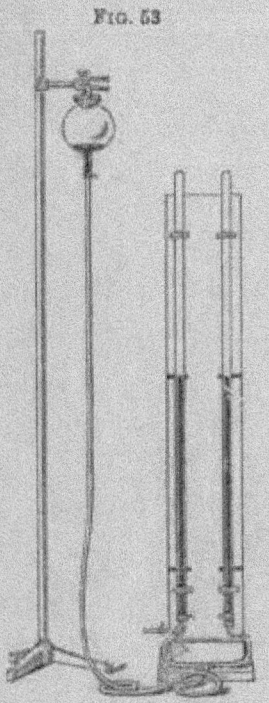

FIG. 53

Both corks should be securely wired into the tubes, which for this purpose should be slightly 'lipped' at the edges.

On raising the reservoir so as to put a pressure of one atmosphere upon the two gases, indicated by the mercury in the air tube rising to the division marked half-way up the scale, it will be seen that the volume of ammonia gas in the other tube is reduced to less than half the original amount owing to additional solution in the water under the increased pressure.

HYDROGEN PEROXIDE. H_2O_2

FORMATION

135. By the action of carbon dioxide upon barium peroxide $BaO_2 + CO_2 + H_2O = BaCO_3 + H_2O_2$. Twenty grams of barium peroxide are mixed with 100 cubic centimetres of distilled water in a beaker, and a rapid stream of carbon dioxide, which has been washed by passing through water, is bubbled through. If the delivery tube reaches to the bottom of the beaker, the mixture will be kept stirred by the bubbles of gas. In a few minutes a piece of potassium iodide and starch paper will show the presence of the hydrogen peroxide.

136. By the action of dilute hydrochloric acid on barium peroxide. $BaO_2 + 2HCl = BaCl_2 + H_2O_2$. Three or four grams of the peroxide are mixed with about 30 cubic centimetres of dilute hydrochloric acid (consisting of 1 part strong acid and 12 parts water). The solution, on filtering or allowing to settle, will give the blue colour with starch paper.

137. By the combustion of a jet of hydrogen gas. About 20 cubic centimetres of distilled water are placed in a platinum

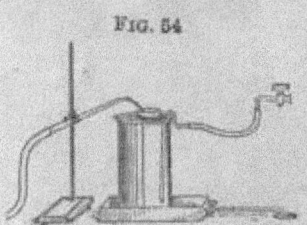

FIG. 54

dish, which is kept cold by being floated in a vessel of running water, as shown in the figure. A jet of burning hydrogen (delivered from an ordinary mouth blow-pipe) is made to impinge upon the surface of the water in the dish in such a way as to ripple up the surface. The jet should be held by a clamp which is standing upon a levelling block (see Experiment 85) so as to admit of nice adjustment. In about 10 minutes the water will contain enough hydrogen peroxide to be detected by the chromic acid test.

138. By the action of dilute hydrochloric acid on sodium peroxide. $Na_2O_2 + 2HCl = 2NaCl + H_2O_2$. To prepare small quantities of hydrogen peroxide for purposes of illustration, the following is a convenient and ready method. About 40 cubic centimetres of dilute hydrochloric acid (consisting of one part of strong acid to ten parts of water) are placed in a small flask, and cooled to 0°C in broken ice. The cooling may be expedited by putting a few fragments of ice into the dilute acid itself. Sodium peroxide is then gradually added in small quantities at a time, avoiding as far as possible any rise of temperature of the solution.

139. Reactions with hydrogen peroxide. To show its action on potassium iodide and starch. This may be done either by dipping starch paper into the peroxide, or by adding the latter to an emulsion of starch, to which potassium iodide has been added; in either case the blue colour does not appear at once, as the reaction is slow.

A striking way of showing the time required for this reaction is to place upon a clean glass plate two or three separate drops of the solution of potassium iodide and starch, and throw an image of the plate on the screen by means of the horizontal projector (see Lantern Illustrations). A drop of peroxide is delivered from a fine tube upon each of them, and gradually a dark edge appears surrounding the added drop, the colour slowly spreading throughout the mass.

140. To show the action upon potassium dichromate. (The chromic acid test.) To a quantity of diluted hydrogen peroxide a few drops of hydrochloric acid are added; and on the addition of one or two drops of the solution of the dichromate a deep-blue colouration is obtained. On shaking up with ether, the blue compound will be dissolved and will appear in the ethereal layer. The compound being very unstable, the blue is extremely fugitive; but it is less so in ethereal than in aqueous solution. In testing for very small quantities of the peroxide, therefore, as in the result of Experiment 137, it is better to add ether before adding the dichromate.

141. To show the action upon lead sulphide. $PbS + 4H_2O_2 = PbSO_4 + 4H_2O$. Sulphuretted hydrogen water is added to a dilute solution of a lead salt, to precipitate the black lead sulphide. On the addition of hydrogen peroxide, the black precipitate is rapidly converted into the white lead sulphate.

Cartridge paper which has been moistened with a solution of a lead salt, and dried, may be blackened by exposure to sulphuretted hydrogen; and on dipping such black paper into hydrogen peroxide, it is immediately restored to its original white colour.

A piece of oil canvas, or, better still, a rough oil painting in which light and bright colours predominate, may be washed over with sulphuretted hydrogen water until all trace of the picture is lost. On brushing the peroxide over a portion of the surface, the colours of the picture will be seen slowly to reappear, and finally will be restored to almost the same brilliancy as before the treatment.

The same canvas or picture may be blackened and restored a number of times.

142. To show the bleaching action of hydrogen peroxide. The best colour to employ for this purpose is aniline blue (Triphenyl rosaniline), as, owing to its high tinctorial power, only a very small quantity is required. The peroxide is tinted by the addition of a small quantity of a solution of the blue, and then boiled; in a short time the colour is discharged. If a few drops of a solution of caustic soda be added to the peroxide, the action takes place almost immediately.

143. To show the action of nascent hydrogen upon hydrogen peroxide. Some granulated zinc and water are placed in the ordinary hydrogen-generating apparatus, and a few drops of sulphuric acid added so as to cause a very slow evolution of hydrogen. The rate of evolution may be shown by collecting

the gas in a tall glass tube. On adding hydrogen peroxide to the contents of the vessel the evolution of gas will stop, as seen by the cessation of bubbles ascending in the tube and the clarifying of the liquid in the generating bottle.

144. This may also be shown by placing the materials in a cell and projecting an image of the liquid upon the screen. The effect of the peroxide in stopping the evolution of gas will be well seen by the disappearance of the bubbles.

145. A still more striking method for illustrating this property of the peroxide is by electrolysis. A feeble current is passed through a quantity of the peroxide contained in a cell in the microscope, and an image of the wires thrown on the screen. It will be seen that gas is only evolved from the positive electrode. A current reverser should be placed in the circuit.

146. If a strong solution of the peroxide is at hand, it is possible to show the absorption of hydrogen evolved in the electrolysis of dilute acid by means of hydrogen peroxide. A small quantity of water acidulated by a drop of sulphuric acid is placed in the microscope cell, the electrodes reaching nearly to the bottom. This is subjected to electrolysis by as feeble a current as is capable of causing its decomposition. On adding strong peroxide, the evolution of gas from the negative electrode will be seen to stop.

147. To show the action of hydrogen peroxide upon certain metallic oxides. $Ag_2O + H_2O_2 = 2Ag + H_2O + O_2$. A small quantity of silver oxide is added to a few cubic centimetres of the peroxide in a test tube, when a rapid evolution of gas takes place at once, which may be shown to be oxygen by introducing a glowing cedar splint.

Manganese dioxide treated in the same way will yield a similar result. $MnO_2 + H_2O_2 = MnO + H_2O + O_2$.

148. To show the action of hydrogen peroxide upon potassium permanganate. $K_2Mn_2O_8 + 3H_2SO_4 + 5H_2O_2 = K_2SO_4 + 2MnSO_4 + 8H_2O + 5O_2$. To a few cubic centimetres of the peroxide about

half its volume of sulphuric acid is slowly added, the mixture being kept cool. A few grams of potassium permanganate are placed in a small flask, fitted with a dropping funnel and a delivery tube, and the mixture gradually dropped upon it. Rapid effervescence takes place, and the oxygen may be collected at the pneumatic trough.

149. **To show the action of heat on hydrogen peroxide.** A small flask, fitted with a delivery tube, is nearly filled with the peroxide, to which a few drops of caustic soda have been added. On warming the liquid, oxygen is evolved, and may be collected at the trough.

150. **To show the action of mercury upon hydrogen peroxide.** A glass tube is closed at one end, and has a stopcock blown upon the other; the tube may conveniently be of about 40 c.c. capacity. 5 cubic centimetres of mercury are introduced by means of a fine drawn-out funnel tube, which will pass through the stopcock, and the tube filled up with hydrogen peroxide. A short delivery tube is connected to the stopcock. On shaking the tube and then opening the cock a quantity of oxygen will be evolved, which can be collected at the trough by means of the delivery tube. The cock is then closed, and the contents of the tube agitated once more, when a further evolution of gas will take place; in this way a considerable quantity of oxygen may be collected in a few minutes.

151. **To show the action of hydrogen peroxide on sodium hydroxide.** $2NaHO + H_2O_2 = 2H_2O + Na_2O_2$. 10 cubic centimetres of a 20 per cent. solution of sodium hydroxide are placed in a small flask and 4 cubic centimetres of hydrogen peroxide (of the strength known as ' 20 volumes ') are added, and the mixture shaken together. On the addition of about 20 cubic centimetres of absolute alcohol the mass will solidify, with crystals of sodium peroxide, $Na_2O_2.8H_2O$. If hydrogen peroxide be added to a solution of barium hydroxide, the corresponding hydrated barium peroxide, $BaO_2.8H_2O$, is at once precipitated.

152. To show the action of hydrogen peroxide upon bleaching powder. $Ca(OCl)Cl + H_2O_2 = CaCl_2 + H_2O + O_2$. A mixture of bleaching powder and water is placed in a flask, fitted with a delivery tube. Hydrogen peroxide is added, and the cork with its tubes is quickly replaced. A brisk effervescence instantly takes place, and the evolved oxygen may be collected in the usual way.

153. To show the action of hydrogen peroxide upon ammoniacal copper compounds. A quantity of saturated copper sulphate solution, to which an excess of ammonia has been added, is placed in a flask carrying a dropping funnel and delivery tube. Hydrogen peroxide is allowed to flow in, and a rapid evolution of oxygen at once takes place. The gas carries with it some of the ammonia, from which it may be purified, if desired, by passing it through dilute sulphuric acid, or by adding a little acid to the water in the trough.

The copper salt remains unchanged at the end of the operation. It acts as a catalytic agent, and is apparently first reduced to the cuprous state, and then reoxidised by another portion of the hydrogen peroxide.

CHLORINE

154. Preparation from hydrochloric acid and manganese dioxide. $MnO_2 + 4HCl = MnCl_2 + 2H_2O + Cl_2$. A large flask is fitted with a cork, preferably of india-rubber, carrying two tubes; one a long straight tube reaching to within two or three centimetres of the bottom of the flask, and extending about 30 centimetres above the cork; the other an exit tube bent at a right angle.

About 300 cubic centimetres of strong hydrochloric acid are placed in the flask, and about 100 grams of the manganese

dioxide added. On the application of a gentle heat a steady
evolution of chlorine takes place. The gas should be washed
by being passed through a Woulf's bottle containing water.

It will be found convenient to attach a T tube between the
generating flask and the wash bottle, as in fig. 55. This tube
is made to dip into a
cylinder containing a so-
lution of caustic soda, of
depth enough to offer
more resistance to the
gas than is offered to it
by the water in the
Woulf's bottle and any
liquids through which it
may be desired to pass it
in the course of experi-
ment. A screw clamp is
provided upon the caout-
chouc, which attaches the
T tube to the wash bottle. By closing this cock the gas is compelled to pass into the soda, where it will be absorbed.

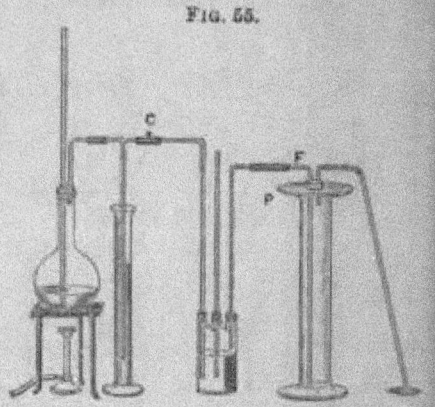

FIG. 55.

If the caoutchouc corks employed are already bored, the
holes should be cleaned out by means of a little bisulphide of
carbon before the glass tubes are inserted, or the tubes may be
made hot enough to slightly burn the rubber; in this way the
tubes will become perfectly firmly attached to the caoutchouc,
and must never be shifted. If the holes are bored at the time
of fitting up the apparatus, the borers should be moistened with
spirit, which not only makes them cut more easily, but leaves
the rubber in a condition to adhere firmly to the glass; by these
devices the apparatus may be made perfectly gas-tight. In
charging the apparatus the acid should be put into the flask
first and the manganese added; for if the acid be poured upon
the manganese the latter is liable to cake upon the flask, which
is then very likely to be fractured: for this reason it is not
advisable to have a funnel fitted to the flask, the long tube

being intended solely as a safety tube. The flask should be supported by a piece of wire gauze upon a tripod in preference, to a sand bath, as the heating is then so much more under control. A slight raising or lowering of the gas flame will at once increase or diminish the rapidity of the evolution of the gas.

For most purposes chlorine is best collected by displacement. A convenient form of apparatus is shown in fig. 56. It consists of a ground glass plate, through a centre hole in which fits a cork carrying two tubes, one passing to the bottom of

Fig. 56. the glass cylinder, and the other ending immediately below the cork. The escaping gases are either passed into caustic soda or discharged into a draught hole. When the cylinder is full, the cock c (fig. 55) is closed, the tube F is disconnected from the wash bottle, and the plate quickly removed and replaced by another previously waxed and ready to hand. If the tube F is made to fit the cork rather easily, so that it may be drawn nearly through before the plate is removed, there will be less time occupied in changing the plates, and consequently less escape of chlorine. A number of cylinders may be connected in series and filled at the same time.

For collecting and keeping gases required for lecture experiments, cylinders with ground edges, covered with waxed ground glass plates, are preferable to stoppered jars or bottles. The best wax for the purpose is resin cerate (procurable from any druggist), which should be of about the consistency of firm lard. If it should be lumpy or too hard it is a good plan to melt it down and add a small quantity of vaseline. The ground glass plates must not be more than ¼ inch larger in diameter than the mouth of the cylinders they are to cover; the wax should be laid on evenly as a narrow ring round the plate with the finger, and the waxed plate pressed firmly down upon the cylinder. Cylinders so filled with chlorine or other gases may be preserved for many months.

G

Chlorine may be collected at the pneumatic trough without any appreciable loss by employing water which has been saturated with common salt.

155. **Preparation from sodium chloride manganese dioxide of sulphuric acid.** $MnO_2 + 2NaCl + 2H_2SO_4 = MnSO_4 + Na_2SO_4 + 2H_2O + Cl_2$. This preparation may be made in an apparatus exactly similar to that used in No. 149. Diluted sulphuric acid (see Preparation of HCl from NaCl) is first introduced into the generating flask, and a mixture of manganese dioxide and common salt in about equal parts is added. On the application of a gentle heat a steady evolution of chlorine takes place.

156. **Preparation from potassium dichromate and hydrochloric acid.** $K_2Cr_2O_7 + 14HCl = 2KCl + Cr_2Cl_6 + 7H_2O + 3Cl_2$. A quantity of the dichromate, coarsely broken up, is placed in a flask fitted with a cork and delivery tube, and a quantity of strong hydrochloric acid poured over it. On applying heat, chlorine is evolved.

157. **Preparation of chlorine by 'Deacon's process.'**

$$(1)\ Cu_2Cl_2 + O = Cu_2OCl_2;$$
$$(2)\ Cu_2OCl_2 + 2HCl = H_2O + 2CuCl_2;$$
$$(3)\ 2CuCl_2 = Cu_2Cl_2 + Cl_2;$$

or

$$(1)\ O + 2HCl + Cu_2Cl_2 = H_2O + 2CuCl_2;$$
$$(2)\ 2CuCl_2 = Cu_2Cl_2 + Cl_2.$$

A porcelain tube is filled with fragments of pumice which have been soaked in a solution of a copper salt, preferably cupric chloride, and dried. The tube is heated to dull redness in a coke or, preferably, a gas furnace. To one end of the tube a delivery tube is attached; the other end is connected to a three-necked Woulf's bottle containing sulphuric acid. Hydrochloric acid is passed into the Woulf's bottle by one opening, and air, or better oxygen, at the other; the rate at which these gases are passing being observed by the rapidity of the bubbles in the Woulf's bottle. Chlorine, mixed with nitrogen if air be used, or with oxygen if that gas be employed, may be collected at the pneumatic trough over brine. The presence of the chlorine in the gas which is so obtained is best shown by introducing a strip of paper coloured with carmine, which will be rapidly bleached.

158. This reaction may be illustrated on a small scale by using a combustion bulb to contain the fragments of pumice (fig. 57). Oxygen and hydrochloric acid passed over in the cold may be shown to undergo no change, by showing that the issuing gas has no action on paper which has been coloured with carmine. On gently warming the bulb, enough chlorine will be produced in a few seconds to show its bleaching power upon the coloured paper.

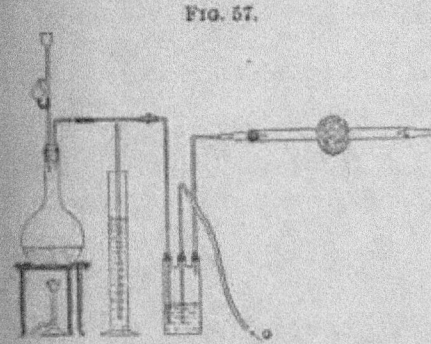

FIG. 57.

The issuing gas may be made to bubble into a solution containing potassium iodide and starch, when instantly upon warming the bulb the blue colour will be manifest.

159. To prepare liquid chlorine. A stout glass tube closed at one end, and bent at a right angle near the middle, is partially filled with the solid hydrate of chlorine (see Experiment 180) and the open end drawn off and sealed up, the tube being kept cold by a freezing mixture. On gently warming one limb of the tube, and cooling the other, a small quantity of liquid chlorine will distil over.

160. A better method for obtaining the pure liquid, consists in passing chlorine from the generating apparatus through a tube, which is cooled by solid carbonic acid and ether. If it be not desired to preserve the sample of liquefied chlorine, the most

convenient tube for its collection is the 'condensing tube' described under Liquefaction of Gases (Experiment 649); but if it be intended to keep the specimen, it is better to collect it in a U tube, made of stout glass, and having its two limbs bent near together so that it will pass into a narrow vessel containing the freezing mixture. The extremities of the U tube should be drawn out, in order that they may be sealed off in a blowpipe flame. The chlorine should be dried by passing through sulphuric acid before entering the cooled tube. A large 'boiling tube,' wrapped round with green baize, is a convenient vessel in which to cool the U tube.

FIG. 58.

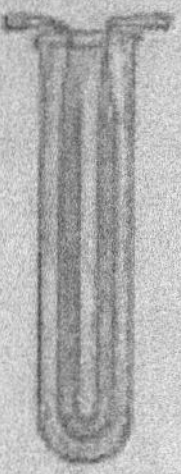

It may be useful to know that the ether vapour, which is evolved from a solution of solid carbon dioxide in ether, is not inflammable, owing to the amount of carbon dioxide which is simultaneously evolved. When sealing the U tube, therefore, it need only be slightly raised in the boiling tube, as it is perfectly safe to direct the flame right across its mouth, and so to seal the tube while it is still immersed in the freezing mixture.

161. To show the action of chlorine upon metals. Antimony. A quantity of finely powdered antimony is tied up in a little muslin bag. This is dusted into a cylinder of chlorine, and each particle as it falls into the gas burns with a bright sparkle, producing fumes of the chloride of antimony. This, and all other experiments with chlorine in which the gas is liable to escape from the jar, must be performed either in a draught cupboard, or over a draught hole upon the table.

162. Brass. A number of leaves of Dutch metal are made into a loose bundle, and attached to the end of a stout wire. On introducing this into chlorine, the bundle of Dutch metal instantly ignites. If the leaves are attached to the end of a spiral of rather fine brass wire, their combustion will be communicated to it, and it will continue to burn in the gas. Sand should be placed in the bottom of the cylinder to prevent its fracture by falling molten metal.

163. This experiment is more strikingly shown by placing a quantity of Dutch metal into a round glass flask fitted with a cork, carrying a glass stopcock. The flask is then exhausted by means of an air-pump, and connected to another flask of larger dimensions and fitted in a similar manner, and which has been filled with chlorine. On opening the cocks, chlorine will rush into the exhausted flask, and the metal in it will be instantly ignited.

164. Sodium. A quantity of sodium is heated in a deflagrating spoon until it begins to burn in the air; it is then introduced into a flask filled with chlorine, when it continues to burn with intense brilliancy. As the deflagrating spoon is liable to melt, a quantity of sand should be placed in the flask.

165. Mercury. A stream of chlorine is passed over mercury contained in a bulb tube. On heating the mercury it inflames and burns with a bright white light.

A wide escape tube should be provided to carry off the mercuric chloride which is produced.

166. To show its action upon phosphorus. A fragment of phosphorus in a deflagrating spoon is introduced, without previous ignition, into a cylinder. The phosphorus spontaneously inflames and burns with a feeble flame.

167. To show its action upon hydrocarbons. Turpentine. A few cubic centimetres of turpentine are heated in a test tube, and poured upon a folded strip of filter paper, which is then introduced into a cylinder of chlorine. Instantaneous inflammation takes place, with the deposition of a large quantity of carbon.

168. Ethylene. A tall cylinder is one-third filled with ethylene at the pneumatic trough, and twice the volume of chlorine added, leaving a little water to facilitate mixing the gases. The

mouth of the cylinder is covered with a plate, and the contents
of the cylinder shaken together for a moment. On applying
a light the mixture burns with a lurid light, depositing carbon,
and forming a cloud of hydrochloric acid.

169. **Coal gas.** A jet of burning coal gas may be lowered
into a cylinder of chlorine, when it will be seen to burn with a
lurid smoky flame, and fumes of hydrochloric acid are evolved.

170. A lighted taper introduced into chlorine behaves in a
similar way.

171. **To show the behaviour of chlorine towards carbon.** A
fragment of charcoal is heated to redness, and plunged into
chlorine. The charcoal will be seen to be at once extinguished.
The charcoal should be held in a loop of stout platinum wire,
attached to the stem of a deflagrating spoon.

172. **To show the bleaching action of chlorine.** A small
quantity of chlorine water may be poured into flasks containing
water which has been tinted with colouring matter, such as
magenta, aniline blue, indigo, carmine.
Paper which has been coloured by such materials may be
dipped into chlorine water, or into cylinders of the gas.
Fabric, dyed with similar colours, may be bleached by being
dipped into chlorine water.

173. **To show the behaviour of dry chlorine towards colouring
matters.** A cylinder of chlorine is dried by a layer of sulphuric
acid being placed in the vessel, and the whole allowed to stand
for a few hours. A dry piece of fabric, dyed 'Turkey red,' is
pinned to a cork which has been cemented to a glass plate, and
is quickly introduced into the dry gas. The colour will not be
discharged, although it will be slightly changed.

174. **To show the behaviour of dry chlorine towards metals.
Brass.** A round flask, of about a litre capacity, is fitted with a
cork carrying a glass stopcock. Into the flask, after being carefully dried, a quantity of phosphoric anhydride is placed, and

about a couple of dozen leaves of Dutch metal are introduced as quickly as possible, and the cork tightly inserted.

A large ordinary flask, of about 3 litres capacity, is fitted as the other. A quantity of the anhydride is placed in it, and the

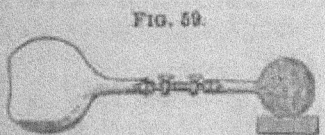

FIG. 59.

flask is filled with chlorine by displacement. The cork is then firmly inserted, and both flasks are left to stand for twenty-four hours.

The stopcocks should be as near to the cork as will admit of turning, and the tubes should be cut off about 1 centimetre from the cock.

The round flask is connected to an air-pump and exhausted, and the two flasks are then connected together with a piece of caoutchouc, the ends of the glass tubes being brought close together (fig. 59). The air thus inclosed between the two cocks, not having been dried, must be as small a quantity as possible. The cock of the chlorine flask is first opened, and then the other. Chlorine rushes into the flask containing the brass, and if the experiment is successful there will be no action, the metal remaining bright and untarnished. The cocks are then closed, and the flasks disconnected. A single drop of water from a pipette may now be introduced into the short tube of the stopcock, and the cock opened; there being less pressure in the flask than outside, the drop will be sucked in, and instantly the brass will combine with the chlorine, bursting into flame.

175. Sodium. A long piece of glass tube, about 2 centimetres bore, is drawn out at one end, and constricted in two places, A', B', as seen in fig. 60, the glass being thickened at the constrictions. A branch tube, E, is blown into the tube at C. A caoutchouc cork, with two holes, is fitted on to this branch tube, and the end of the tube is then drawn out fine, and the fine tip bent at right angles and sealed. The second hole in the cork carries a piece of glass rod, reaching to within about 2 or 3 millimetres of the fine tube. A lump of sodium, cut from an ingot of the metal so as to be as clean as possible, is pushed into A, and the tube is sealed off at a. The length of this part of the tube should be sufficient to prevent moisture from the blowpipe flame from condensing inside. The tube D is now connected to a Sprengel pump, and the apparatus exhausted. As soon as the mercury in the pump begins to click, the tube should be gently warmed by passing a Bunsen flame lightly along it, taking care to warm every part, from A to D and along E. The pump will at once show that there are vapours produced. When the clicking begins again, the tube may be made hot enough to melt the sodium, and the liquid metal should run out of the shell of oxide which coats it, exposing a brilliant sur-

FIG. 60.

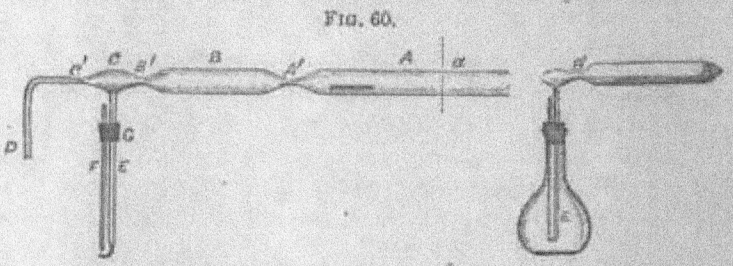

face. When a good vacuum is again obtained the tube is to be sealed off at c' by means of a fine blowpipe flame. The sodium is again melted, and the whole tube made warm by passing the flame along it. The melted metal is then allowed to flow through the constriction A' into B ; the constriction effectually stopping every trace of 'dross,' so that the sodium lies in B as a perfectly clean and brilliant mass. The tube is allowed to cool, and is sealed off at A' as before. Should a fragment of sodium remain in the constriction A', it is easily distilled away by gently warming the spot, before sealing up. If desired, the metal may be again melted and made to take up any required position in the tube.

A flask of about ¾ litre capacity, the neck of which will fit the cork G, has phosphoric anhydride placed in it, and is filled with chlorine by displacement. The cork G is firmly introduced, and the apparatus allowed to stand for at least twenty-four hours (fig. 60). At the end of this time the fine point of tube E is broken off by depressing the glass rod upon it. Chlorine instantly rushes into the vacuous tube containing the bright metal, and it will be seen to exert no action upon it. The tube is then sealed off at E′, there being still a slight reduction of pressure within the apparatus, and the specimen will retain its bright metallic lustre indefinitely.

176. **To show the combination of chlorine with hydrogen.** Two small stout glass cylinders may be filled ; one with chlorine, and the other with hydrogen. These are placed mouth to mouth, the plates withdrawn, and, after being once or twice inverted to insure the mixture of the gases, a lighted taper is applied to each. A sharp explosion follows, and clouds of hydrochloric acid make their appearance.

177. A jet of burning hydrogen may be lowered into a cylinder of chlorine. The hydrogen burns with a pale yellowish flame, and hydrochloric acid fumes are seen escaping from the mouth of the vessel. (See also Experiment 182.)

178. **To show the effect of light in causing the combination of chlorine and hydrogen.** The plan frequently described for performing this experiment, viz. of filling a bottle with a mixture of chlorine and hydrogen and throwing it from an

FIG. 61.

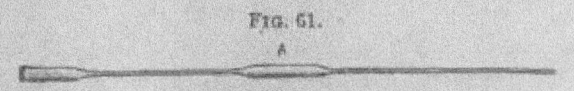

open window into the bright sunshine, is for many obvious reasons unsuitable for a lecture experiment. The best method is to fill small glass bulbs with the mixed gases, and expose them to suitable artificial light.

To make the bulbs, soft glass tube, either soda or lead glass (about 3 millimetres bore and 1 millimetre thickness of wall),

is drawn out, as shown in fig. 61. The capillary tubes should
be as fine as will allow of a tolerably free passage for the gases; if
too fine, the bulb will be difficult to fill, and if too coarse the after-
operation of sealing up cannot be performed without exploding
the gases. The portion of the original tube, upon which the bulb
is to be blown (A, fig. 61) may be about $2\frac{1}{2}$ to 3 centimetres
long. The bulb when blown should be about the size of a hen's
egg, and will be quite thin.

The bulbs are filled with the gases evolved by the electrolysis of hydrochloric acid.

A small apparatus is fitted up similarly to the one described
for the electrolysis of water; the electrodes in this case being
either of stout platinum or small rods of carbon (such as are
used for 'arc' lights); the action of the chlorine upon the
platinum is so slight that it is of practically no importance.
If carbon electrodes are used, they may be attached to short
pieces of platinum wire, which are fused into glass tubes in
order to make a mercury contact with the battery. The
apparatus should be filled to within about $2\frac{1}{2}$ centimetres from
the neck with hydrochloric acid of specific gravity 1·10 (approximately obtained by mixing 18 volumes of strong acid with 10
volumes of water), and a current from five or six Grove's cells
passed through. It is necessary that the temperature of the
acid should not be allowed to rise, and to prevent this the
apparatus must be stood in a large vessel of cold water, or a
vessel through which cold water is circulating. When the
current has been passing for ten or fifteen minutes, the liquid
may be considered to be saturated with chlorine, and a bulb
may be filled. A piece of caoutchouc tube 15 to 20 centimetres
long is attached to the delivery tube, and the other end connected to the bulb to be filled, and the gases direct from the
electrolytic cell passed through (fig. 62). *The gases should not
be dried by being passed through sulphuric acid.* With the
strength of battery indicated, and if the capillary tubes of the
bulb are not too fine, five minutes is ample time to allow for the
filling of a bulb. A strip of carmine paper may be brought to

the point where the gases escape, to ascertain that chlorine is actually passing out. At the end of about five minutes the caoutchouc tube is disconnected from the bulb, and the bulb is

Fig. 62.

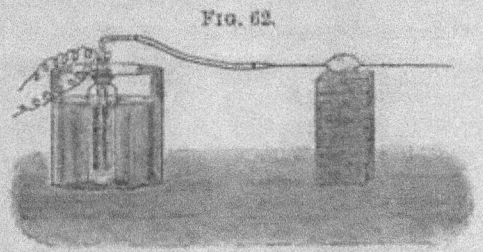

sealed up before the blowpipe. If the capillary tubes are about the right size, this operation may be done without any risk of the mixture being ignited. A small fine blowpipe flame must be used, and the little tube passed backwards and forwards across the flame, a gentle pull being exerted on the tube all the time ; in this way the glass is never heated higher than is absolutely necessary to soften it. The sealing should be made at a point

Fig. 63.

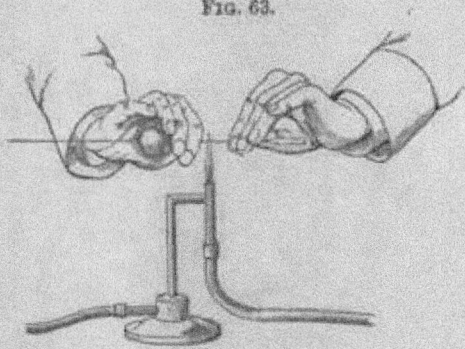

about 5 centimetres distant from the bulb. In order to protect the face and eyes in the event of the mixture exploding during this operation, the bulb should be held in such a way that the palm of the hand forms a screen (fig. 63). When these bulbs burst the glass is reduced to such fine dust that the hand feels absolutely nothing from the explosion. The operator may use a glove upon the hand if desired; but a gloved hand will be found to interfere with the delicacy of manipulation required for these thin bulbs and fine tubes. The entire operation of filling and sealing these bulbs must be carried on by gas-light, or in very diffused daylight. The first bulb that is filled should be tested by being exposed to the light of burning magnesium; if the mixture explodes, it may be presumed that the next bulb, filled in the same way, will do the same. It is well to fill a number of these bulbs at one time, and after, say, a dozen or two have been made, the last should also be tested. They may be conveniently preserved in cardboard egg boxes, and if placed in the dark will keep indefinitely.

The bulbs may be exploded by a variety of different lights.

(a) *Burning magnesium ribbon.*—The bulb is placed in a suitable support (a bent wire loop, standing in the chimney

FIG 63A.

of a Bunsen lamp answers very well, fig. 63A), and surrounded either with a bell-jar or a large beaker, or two sheets of glass hinged together by a strip of cloth glued down the edges. A strip of burning magnesium brought near to the glass screen will cause the combination of the gases and the explosion of the mixture.

(b) *Magnesium 'flash' light.*—A bulb supported and screened as before, and exposed to the light of the magnesium 'flash' used by photographers, will be exploded. An arrangement to produce this flash is described in Experiment 50, page 27.

(c) *By the combustion of a mixture of nitric acid and carbon disulphide.* (See Nitric Oxide, Exp. 319.)

(d) *By the electric light.*—The most striking way of performing this experiment is to hang the bulb, by means of a thread, in the path of the beam from the arc lamp, and project an image of it upon the screen by placing in front of the condensing lens a sheet of red glass. The position in which to place the bulb, and the adjustment of the focus should be previously arranged with an empty bulb. The charged bulb is then put in its place and its image thrown on the screen; on withdrawing the red glass the bulb will be seen to give a kick, and instantly to disappear at the moment of the explosion. The bulb must not be exposed to the red light for more than a few seconds, so that the coloured glass must be removed almost immediately after showing the bulb on the screen.

179. To show that no change in volume follows the combination of chlorine with hydrogen. A glass tube about 36 centimetres long, and rather thick in the walls, has a stopcock blown upon each end. Two platinum wires are sealed into the glass. This tube is filled with the mixed gas from the electrolysis apparatus described above. By dipping one end under mercury and opening the cock for an instant it may be shown that the gases in the tube are at the ordinary pressure. An electric spark is then passed through the gases (and for this it is well to cover the tube with a wire gauze jacket, as it is liable to burst with the force of the explosion), and the cock again opened under mercury, when it will be seen that the pressure inside the tube has remained the same. On opening the cock under water the liquid will be seen to rise in the tube, and completely absorb the hydrochloric acid that has been formed.

180. To show the formation of the solid hydrate of chlorine. $Cl_2, 10H_2O$. A stream of chlorine is passed through water contained in a flask which is kept cold by ice to which a sprinkle of salt has been added. After a short time crystals of the hydrate deposit, and the mass becomes semi-solid.

HYDROCHLORIC ACID. HCl.

181. Preparation from sodium chloride. $NaCl + H_2SO_4 = NaHSO_4 + HCl$. A flask of about $1\frac{1}{2}$ litre capacity, fitted with a mercury safety funnel and delivery tube, has a quantity of common salt placed in it (100 to 150 grams), and is then about half filled with sulphuric acid, previously diluted. The best degree of dilution is obtained by mixing eleven volumes of the strong acid with eight of water, and allowing the mixture to cool. It is convenient to keep a supply of acid of this strength. This acid when poured upon the salt causes no evolution of gas in the cold, but a steady stream is obtained on the application of a gentle heat, and the evolution of gas can be regulated to a nicety by raising or lowering the gas. The gas may be passed through a Woulf's bottle containing sulphuric acid, and can be collected either over mercury or by displacement (fig. 64). It will be found convenient to attach a three-way tube to the generating flask, the long limb of which dips into water contained in a cylinder. When the gas is to be collected over mercury, or has to overcome any resistance in its passage through any apparatus it may be required to traverse, a layer of mercury should be placed in the bottom of the cylinder, and by raising the cylinder upon a block the tube may be made to dip beneath the metal. The quantity of mercury in the funnel should be so arranged that less pressure is required to draw air into the flask through the mercury than is required to raise the acid in the Woulf's

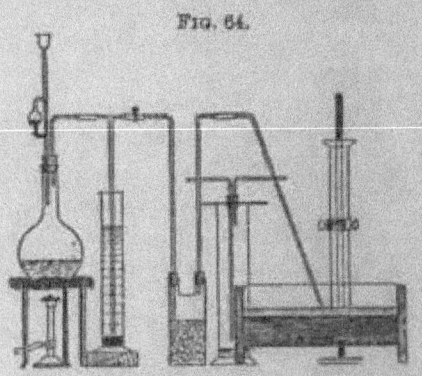

Fig. 64.

bottle to the top of the tube which conveys the gas; so that when the flask cools the strong acid is not drawn over.

182. Formation of hydrochloric acid by synthesis. A slow stream of hydrogen is sent through a small Wurtz flask (about 300 c.c. capacity) which is held in an inverted position in a clamp; the hydrogen being introduced by the side tube. A

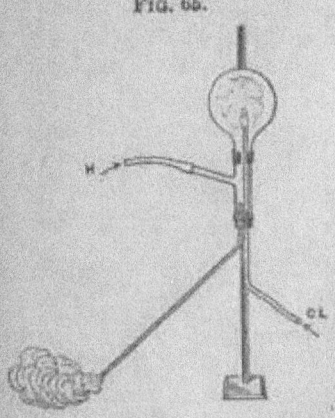

Fig. 65.

cork which fits the flask is provided with a straight glass tube, reaching just up to the body of the flask, and a second tube passing just through the cork and bent at an obtuse angle. The straight tube is connected by means of caoutchouc tube to a chlorine generator direct (without a wash bottle). When the Wurtz flask is full of hydrogen the gas is ignited at the mouth of the flask, the cork is then quickly introduced; as the end of the tube conveying the chlorine passes the flame of hydrogen the chlorine ignites and continues burning in the hydrogen; the excess of hydrogen, along with torrents of fumes of hydrochloric acid, will issue from the end of the bent tube. If the issuing gas be passed into water the acid nature of the product may be demonstrated.

183. To show the electrolysis of hydrochloric acid and the collection of the mixed gases. The apparatus described under Chlorine (Exp. 178) is employed, and the mixed gases collected in a stout glass cylinder over saturated brine in the pneumatic trough.

184. To show equal volumes of chlorine and hydrogen in the mixed gases. A glass tube about 60 centimetres long has a small stoppered funnel blown upon one end (fig. 66). The tube is divided into two equal parts by means of two india-rubber rings slipped over it (such a ring is conveniently made by cutting a short section from a piece of caoutchouc tube). The tube is filled down to the second ring with the mixed gases evolved by electrolysis, and collected over brine. A solution of potassium iodide is introduced into the tube by means of the stoppered funnel, care being taken to avoid allowing air to enter. The chlorine attacks the potassium iodide, liberating iodine, and the hydrogen is left, and will be found to occupy the space down to the first division, viz. one-half the original volume.

FIG. 66

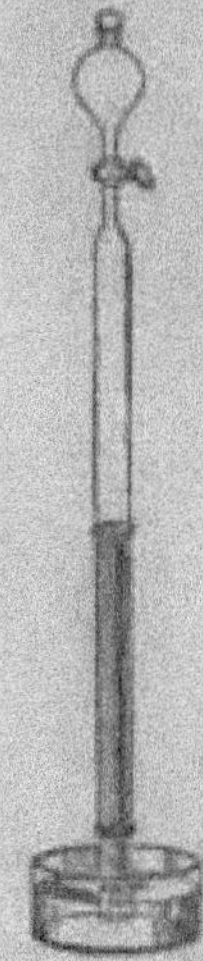

185. To show the electrolysis of hydrochloric acid and the collection of equal volumes of the two gases. A convenient form of apparatus is seen in fig. 67, in which the two tubes are connected by a cross tube out of which a small branch tube rises which is connected to a small reservoir (conveniently a dropping funnel) by means of a caoutchouc tube. The apparatus is supported by a clamp upon a retort stand, the reservoir being held in a ring. The platinum wires carrying the electrodes are fused into short glass tubes which can be fixed into the wider tubes by short pieces of caoutchouc.

FIG. 67.

The apparatus is filled with hydrochloric acid (sp. gr. 1·1) and the reservoir raised until the liquid in the two tubes and the funnel stands about 1 centimetre below the top of the tubes. A current from 7 to 10 Grove's cells is passed through the liquid, the two cocks being open, until the liquid in the tube attached to the positive wire of the battery is saturated with chlorine. It is not difficult to see by the ascending bubbles when this point is reached; it will require from 20 to 40 minutes, depending upon the strength of current used. The cocks are closed simultaneously, and the two tubes will be found to fill with gas in practically equal volumes.

186. To show the behaviour of gaseous hydrochloric acid towards ordinary combustibles. A burning taper, or jet of coal-gas, when introduced into the gas is at once extinguished.

187. To show the combustion of potassium in gaseous hydrochloric acid. A fragment of potassium is placed in a combustion bulb, which is attached to the apparatus, evolving hydrochloric acid. The gas should be dried by being passed through a Woulf's bottle containing fragments of pumice moistened with sulphuric acid, instead of being allowed to bubble through acid, as, in order that the issuing hydrogen may be inflamed, the current of gas must not be intermittent. On heating the potassium it inflames and continues its combustion in the stream of gas. The liberated hydrogen may be ignited as it issues from the bulb tube.

188. To collect hydrogen from the decomposition of gaseous hydrochloric acid by means of sodium. A few fragments of sodium are dissolved in mercury by pressing the sodium in small pieces under mercury contained in a mortar; each piece of sodium as it is added dissolves in the mercury with some energy. The amalgam so obtained is placed in a small flask fitted with a cork and two tubes; one is connected with the hydrochloric acid generator, to the other is attached a delivery tube to convey the gas to the pneumatic trough. On passing the gas, and gently shaking the amalgam in order to constantly expose a fresh surface, a rapid stream of hydrogen will be evolved.

189. To show the volumetric composition of hydrochloric acid by means of sodium amalgam. A measured volume of dry gaseous hydrochloric acid is introduced into the closed limb of a U-shaped eudiometer, and the mercury levelled in both limbs. The volume occupied by the gas is indicated by an india-rubber ring upon the tube, and a second ring marks half the volume. The open limb is then filled up with sodium amalgam, and on being closed with the thumb the gas can be decanted into it, and made to bubble once or twice through the amalgam. It is

H

finally returned to the graduated limb, and on again adjusting
the levels of the mercury in the two tubes it will be found that
the volume of the gas is reduced to one
half. To show that the remaining gas is
hydrogen, it may once more be transferred to the open limb by first filling it
with mercury and closing it with the
thumb, and then inflamed.

FIG. 67A.

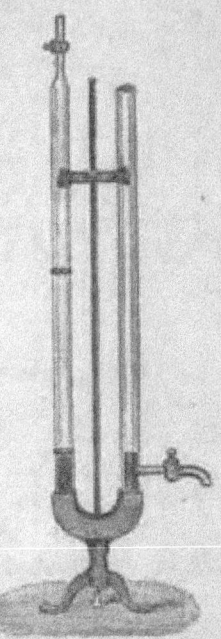

It is more convenient to construct
the apparatus with a stopcock blown
upon the closed limb, as shown in
fig. 67A. The hydrogen may then be
driven out and ignited as it issues.
This experiment only proves that the
volume of hydrogen in the compound is
equal to one half the volume of the
hydrochloric acid, but gives no informa-
tion as to the volume of the chlorine.

190. To decompose gaseous hydrochloric
acid by means of manganese dioxide. A
piece of glass tube about 20 centimetres long is filled with coarsely
powdered manganese dioxide, and a
stream of hydrochloric acid gas is passed over it. On the application of a gentle heat by simply passing a Bunsen flame once
or twice along the tube, chlorine will be evolved, which may be
collected over brine; the undecomposed hydrochloric acid being
absorbed in the water.

191. To decompose gaseous hydrochloric acid by electric
sparks. An apparatus similar to that employed for the decomposition of steam by electric sparks (Water, Exp. 82) may be
used, the materials for generating the gas being placed in the
flask. When the sparks are passed, small quantities of gas are
carried over in the stream of hydrochloric acid, which may be

collected over salt water and shown to contain chlorine, and to explode on ignition.

192. The solubility of hydrochloric acid in water may be illustrated by opening a cylinder of the gas beneath water, which will instantly rush into the cylinder as into a vacuum. The cylinder may be filled by displacement, and should be closed with a thick glass plate. The ordinary thin ground-glass covers should not be employed, as they are certain to be fractured by the pressure at the moment water enters, and the hand of the operator is in danger of being seriously cut; on no account should a stoppered cylinder be used, as upon the entrance of the first drops of water the stopper will be immovably fixed.

193. On a larger scale this experiment may be strikingly made by filling a large bolt-head flask with the gas. The flask is fitted with a caoutchouc cork carrying two tubes 7 or 8 millimetres bore; one reaches to within a short distance of the bottom, while the second is a short tube ending 1 or 2 centimetres through the cork. The ends of these tubes outside are slightly opened out, to admit of their being securely plugged with caoutchouc stoppers. The flask, supported neck upwards, is filled with the gas by displacement, the gas being delivered by the long tube. The issuing gas is passed into water until the air is entirely displaced, which is ascertained by the complete absorption of the bubbles, and the two tubes are then closed. The flask is supported by a tall tripod in a large trough or basin of water, which may be coloured blue with litmus, and the cork of the long tube withdrawn. The water then begins slowly to rise in the tube until it reaches the top, and when the first few drops have entered the flask, and by their absorption of the gas produced a partial vacuum, the water is forced up the tube in the form of a fountain which continues until the flask is filled.

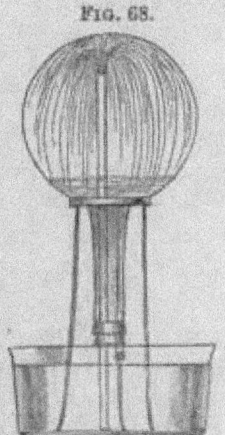

Fig. 68.

OXIDES AND ACIDS OF CHLORINE

CHLORINE MONOXIDE. *Chlorous oxide.* Cl_2O.

194. Preparation by the action of chlorine upon mercuric oxide. $2HgO + 2Cl_2 = HgO,HgCl_2 + Cl_2O$. Chlorine, dried by being passed through sulphuric acid, is slowly passed over dry precipitated mercuric oxide (the red crystalline oxide will not answer) contained in a glass tube. A brownish-coloured oxychloride $HgO.HgCl_2$ is formed in the tube, and the yellow chlorous oxide passes on, and may be collected by displacement. If passed through a U tube contained in a freezing mixture it condenses to an orange-yellow liquid, but as the liquid is highly explosive it is unadvisable to collect it.

CHLORINE PEROXIDE. ClO_2.

195. Preparation by the action of sulphuric acid upon potassium chlorate. $3KClO_3 + 2H_2SO_4 = KClO_4 + 2HKSO_4 + H_2O + 2ClO_2$. About 5 cubic centimetres of strong sulphuric acid are placed in a short stout glass cylinder, covered with a small piece of cardboard or a mica plate: a glass plate should not be used, as the gas is liable to explode during its preparation. Finely powdered and dry potassium chlorate is added in small quantities at a time. The deep yellow gas is evolved, and gradually displaces the air from the cylinder.

196. Preparation by the action of oxalic acid upon potassium chlorate. $2KClO_3 + 2H_2C_2O_4 = K_2C_2O_4 + 2H_2O + 2CO_2 + 2ClO_2$. 18 grams of powdered oxalic acid and 24 grams of powdered potassium chlorate are mixed in a dry flask, and heated to 70°C. in a water-bath. A mixture of the peroxide with carbon dioxide in equal volumes is evolved. The gas may be passed into water, and a solution of the peroxide obtained.

197. **Preparation by the action of hydrochloric acid upon potassium chlorate.** $4KClO_3 + 12HCl = 4KCl + 6H_2O + 9Cl + 3ClO_2$. Hydrochloric acid is poured upon a quantity of crystals of potassium chlorate contained in a flask, and the mixture gently warmed. A mixture of chlorine peroxide with chlorine is evolved, which may be collected by displacement.

(N.B. This mixture of gases was formerly regarded as a distinct compound of chlorine and oxygen, and received the name of euchlorine.)

198. **To show the effect of heat upon chlorine peroxide.** A hot glass rod or copper wire is dipped into a cylinder of the gas obtained by Exp. 195. The gas explodes more or less violently in proportion as it is free from air.

199. **To show the action of chlorine peroxide upon phosphorus.** A few crystals of potassium chlorate along with two or three fragments of phosphorus are placed at the bottom of a cylinder of water. Strong sulphuric acid is poured upon the mixture by means of a thistle funnel reaching to the bottom of the vessel. The peroxide liberated by the action of the acid upon the chlorate at once attacks the phosphorus, which burns in the gas beneath the water. It is well to use a thistle funnel the stem of which has a very small bore.

200. **To show the action of chlorine peroxide upon sugar.** Potassium chlorate and white sugar, both finely powdered, are carefully mixed with a paper-knife or spatula. A small heap of the mixture placed upon a block of wood is touched with a glass rod which has been dipped into strong sulphuric acid. The mixture is at once inflamed.

201. **To show the action of chlorine peroxide upon sulphuretted hydrogen.** A glass jet from which sulphuretted hydrogen is issuing is lowered into a cylinder of gas, obtained by the action of hydrochloric acid upon potassium chlorate; the sulphuretted hydrogen will spontaneously inflame and continue to burn in the gas.

HYPOCHLOROUS ACID. HClO.

202. Preparation by passing chlorine through water in which precipitated oxide of mercury is suspended. $HgO + H_2O + 2Cl_2 = HgCl_2 + 2HClO$. The mixture of water and the yellow oxide of mercury is contained in a flask fitted with a cork and two tubes. The gas is bubbled through the water, the contents of the flask being frequently shaken.

The liquid may be allowed to settle and the clear solution decanted off. This specimen may be preserved.

203. Preparation by the action of nitric acid on bleaching powder. $2Ca(OCl)Cl + 2HNO_3 = CaCl_2 + Ca(NO_3)_2 + 2HClO$. Dilute nitric acid (1 vol. of acid to 15 of water) is slowly added to a clear solution of bleaching powder in water, with frequent stirring; if performed with care and an excess of acid avoided, no chlorine is evolved by this addition of acid. The mixture may then be distilled from a retort, the distillate being dilute hypochlorous acid.

204. By the action of carbonic acid upon bleaching powder. $2Ca(OCl)Cl + H_2CO_3 = CaCl_2 + CaCO_3 + 2HClO$. A stream of carbon dioxide is passed through a mixture of bleaching powder and water for five or ten minutes. On filtering the liquid a yellow solution is obtained, which may be shown to possess bleaching properties.

The same reaction may also be shown by moistening a piece of litmus paper with a solution of bleaching powder and then breathing upon the paper. The paper, which is at first rendered blue, is gradually bleached by the liberated hypochlorous acid.

205. Preparation by the action of chlorine upon calcium carbonate suspended in water. $CaCO_3 + H_2O + 2Cl_2 = CaCl_2 + CO_2 + 2HClO$. A stream of chlorine is passed through water containing precipitated calcium carbonate in suspension in an arrangement similar to that described in Exp. 202. Carbon

dioxide is evolved, and a solution containing hypochlorous acid and calcium chloride is obtained.

206. The bleaching action of hypochlorous acid may be shown by dipping a strip of paper coloured with carmine into the acid obtained by Exp. 202.

207. To show the action of silver oxide upon hypochlorous acid. $Ag_2O + 2HClO = 2AgCl + H_2O + O_2$. Three or four grams of precipitated silver oxide are placed in a small flask fitted with a cork and delivery tube, and the flask filled to the neck with hypochlorous acid. On the application of a gentle heat oxygen gas is evolved, and may be collected at the pneumatic trough.

CHLORIC ACID. $HClO_3$.

208. Preparation from barium chlorate. $Ba(ClO_3)_2 + H_2SO_4 = BaSO_4 + 2HClO_3$. Dilute sulphuric acid is added to a solution of barium chlorate until the barium is all precipitated as sulphate ; the solution is allowed to settle, and the clear liquid may be concentrated in vacuo over sulphuric acid.

209. To show the bleaching action of chloric acid. The dilute acid made as above, but without being concentrated, may be added to a small quantity of water in a test tube tinted with aniline blue. On gently warming the liquid, the colour will be discharged.

210. To show the action upon organic matter. An impure acid (probably containing perchloric acid) made be made by adding to 5 or 6 cubic centimetres of sulphuric acid (11 vols. of acid to 8 vols. of water), contained in a small beaker, about the same number of grams of powdered strontium chlorate in small quantities. On pouring a drop or two of the liquid so obtained upon filter paper which has been moistened with turpentine, the turpentine will be at once inflamed.

PERCHLORIC ACID. $HClO_4$.

211. Preparation from potassium perchlorate. $2KClO_4 + H_2SO_4 = K_2SO_4 + 2HClO_4$. Pure potassium perchlorate mixed with about five times its weight of strong sulphuric acid, is gently distilled in a small retort. A fuming liquid distils over, which may be collected in a test tube.

As the distillation proceeds, a portion of the perchloric acid is decomposed into lower oxides of chlorine and water. The water thus formed combines with the first portions of the distillate, yielding a crystalline hydrate of the acid, having the composition $HClO_4.H_2O$. The operation, therefore, should not be pushed too far; and for the following experiments the first drops of the distillate are the best suited.

Care must be taken that the perchlorate is free from chlorate, and it is well to surround the apparatus with a glass screen.

212. To show the action of perchloric acid upon water. A drop may be poured into water, or a small beaker containing water may be made to receive the acid as it drops from the end of the retort; each drop as it falls into the water combines with it with a hissing noise.

213. To show the action of perchloric acid upon organic matter. A piece of wood is charred by being heated for a few moments in a flame, and a drop of the acid allowed to fall upon it; the drop decomposes with a sharp explosion. Charred filter paper will produce the same effect.

214. To show the bleaching action of perchloric acid. A small quantity of the solution obtained by dissolving the acid in water may be added to a solution of aniline blue, when the colour will be instantly destroyed.

_____

BROMINE

215. Preparation from potassium bromide. $MnO_2 + 2KBr + 2H_2SO_4 = MnSO_4 + K_2SO_4 + 2H_2O + Br_2$. A quantity of manganese dioxide is mixed with sulphuric acid, previously diluted in the proportion of 11 volumes of acid to 8 volumes of water, and the mixture poured into a retort. Crystals of potassium bromide are added, and the mixture gently warmed. The receiver should be surrounded with broken ice.

216. To show the extraction of bromine by means of chlorine. Chlorine water is added to a solution of potassium bromide in a stoppered cylinder; ether is then added and the mixture shaken together; the ethereal solution of bromine rises to the surface, and may be separated by a separating funnel.

217. To show the action of bromine upon metals. A small fragment of antimony may be dropped upon bromine contained in a test tube; the metal inflames as it floats upon the surface of the bromine, burning with considerable energy.

218. To show the difference between the behaviour of dry and wet bromine towards metals. A small quantity of bromine is put into a test tube and a few crumpled fragments of copper foil are dropped into it. No action takes place, but if a few drops of water are added the bromine at once attacks the copper with considerable energy, forming cuprous bromide.

219. To show the bleaching action of bromine. A solution of bromine in water is added to water tinted with carmine, or litmus, when the colour will be discharged.

220. To prepare the hydrate of bromine. $Br_2, 10H_2O$. Water which is saturated with bromine is placed in a test tube, and cooled by immersion in ice to which a sprinkle of salt has been added; in a few moments the solid compound crystallises out.

HYDROBROMIC ACID. HBr

221. Formation by burning hydrogen in bromine vapour. A small quantity of bromine is poured into a glass cylinder, and allowed to vapourise. A jet of burning hydrogen lowered into the cylinder continues its combustion, and clouds of hydrobromic acid are evolved.

222. By the action of water on phosphorus bromide. $P + 5Br + 4H_2O = H_3PO_4 + 5HBr$. A flask fitted with a

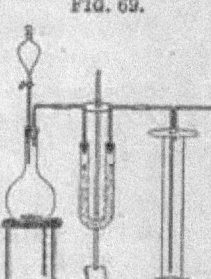

Fig. 69. dropping funnel and delivery tube has placed in it 20 grams of red phosphorus and 40 c.c. of water. Bromine is allowed to drop upon this mixture from the funnel. The addition of the first drops of bromine is attended with the production of a lambent flame, but as the air becomes displaced the action goes on without this appearance of combustion. In order to remove from the hydrobromic acid the bromine vapour which is carried over with it, the gas may be passed through a U tube containing red phosphorus.

Fig. 69 represents a convenient apparatus for preparing and collecting this gas, which owing to its solubility in water must be collected either over mercury or by displacement.

223. By the action of bromine upon melted paraffin. A quantity of solid white paraffin is placed in a small flask fitted with a dropping funnel and delivery tube. The paraffin is melted and bromine slowly dropped upon it. An evolution of hydrobromic acid takes place, and the gas may be freed from bromine vapour, and collected as described above. A similar reaction takes place if turpentine is substituted for the paraffin.

224. To prepare an aqueous solution of hybrobromic acid. The gas prepared by Experiment No. 222 is passed directly into water contained in a Woulf's bottle, but as, from the nature of this method for the preparation, the gas is not evolved in a steady stream, but intermittently as the bromine is introduced, some device must be adopted to prevent the water from passing back into the flask.

A convenient plan is to introduce a safety tube between the U tube containing the red phosphorus and the Woulf's bottle Fig. 70 shows such a safety tube, consisting of an ordinary 'thistle' funnel with a small bulb blown upon the stem, and bent up in the form of a U, a short piece of glass being blown

FIG. 70.

FIG. 71.

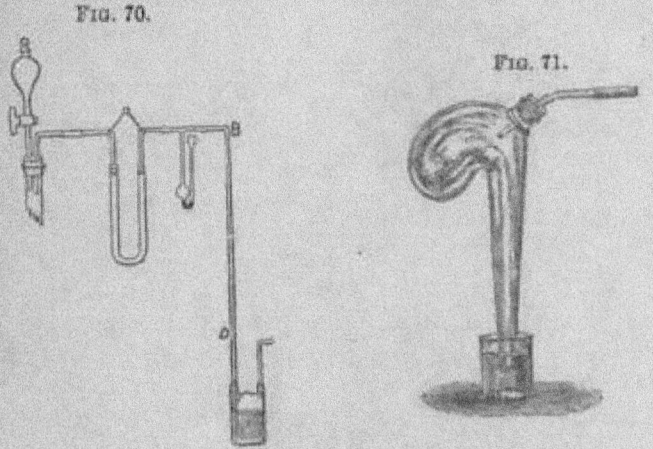

upon the end to form a T. A few cubic centimetres of mercury are placed in the bend of the tube. Gas cannot pass out until the pressure is great enough to lift the mercury up the straight stem into the thistle head, but on a slight reduction of the internal pressure the mercury is drawn up into the little bulb, and air passes in. The tube D must be of such a length that the pressure required to raise a column of water up to the bend B shall be greater than that necessary to allow air to bubble past the mercury in the safety tube.

Another method is to pass the gas through a tubulated retort, arranged as shown in fig. 71. The gas enters through the tubulure, and the stem of the retort dips three or four

centimetres beneath the surface of water; in the event of the
water being sucked back, it rises into the neck of the retort,
and falls into the bulb, allowing air to pass up.

225. A solution of hydrobromic acid may be conveniently
made by the following synthetical process, viz. by passing a
mixture of hydrogen and bromine vapour over a red-hot spiral
of platinum wire. For this purpose a glass tube, about 18 centimetres long and 2 centimetres wide, is fitted at each end with
an india-rubber cork, carrying a short straight piece of narrow
tube. Through each cork a stout iron wire is passed. These
wires are joined together by a short spiral of platinum wire,
about 20 millimetres long. To put the apparatus together, one
of the corks is inserted, with the wires attached (see fig. 72),

FIG. 72.

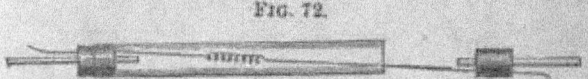

and the projecting wire pushed into a pin-hole previously
made in the other cork. When the wire is through the cork,
the projecting end can be held by a pair of pliers, and the
cork drawn along to its place in the glass tube. One end of
the apparatus is connected to a small wash-bottle containing a
quantity of bromine, through which a stream of hydrogen can
be made to bubble. The other end may be attached to a
Woulf's bottle containing water, or simply to a delivery tube
dipping into a beaker of water. Hydrogen is first slowly passed
through the apparatus for one or two minutes, to sweep out the
air, and the little platinum spiral is then heated to bright redness by means of an electric current. Complete combination of
the hydrogen and bromine at once takes place in contact with
the hot wire, and perfectly colourless hydrobromic acid gas
passes out from the tube, along with an excess of hydrogen.
The supply of bromine vapour may be regulated by standing
the bottle containing the bromine in warm water. About
60°C. is a convenient temperature, and it will be seen

that under these circumstances there is very little hydrogen escaping. So long as even a slight excess of hydrogen is passing (which is readily seen by the escape of bubbles through the water in the absorbing vessel), the issuing hydrobromic acid gas will re-

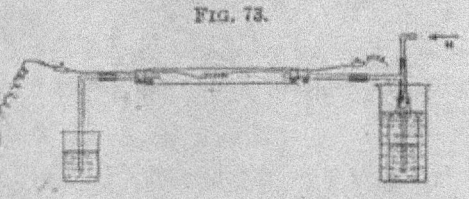

FIG. 73.

main perfectly colourless and free from bromine vapour; it is therefore not necessary to scrub the gas through a tube containing red phosphorus. When the reaction is going pretty rapidly, a lambent flame occasionally makes its appearance in the neighbourhood of the platinum wire, but it shows no tendency to strike back along the narrow tube into the bromine bottle. If desired, however, the additional precaution may be taken of plugging this fine tube with a little piece of glass wool, which renders any inconvenience from this cause quite impossible.

226. Hydrobromic acid in dilute solution may be prepared by passing a stream of sulphuretted hydrogen through bromine water, and filtering the solution to remove the precipitated sulphur.

227. To show the solubility of hydrobromic acid in water. A cylinder is filled with the gas by displacement, and covered with a thick glass plate. On removing the plate under water, the water will rush into the cylinder and completely fill it.

228. To show the action of chlorine upon hydrobromic acid. Two cylinders, one filled with gaseous hydrobromic acid gas, and the other with chlorine, are placed mouth to mouth, and the plates withdrawn. When the gases mix, the brown vapour of bromine makes its appearance, and on inverting the cylinders, in order to mix the gases thoroughly, the action will be complete.

IODINE

229. Preparation from potassium iodide. $2KI + MnO_2 + 2H_2SO_4 = K_2SO_4 + MnSO_4 + 2H_2O + I_2$. This experiment is conducted in the manner described for the preparation of bromine. In this case it is not necessary to cool the receiver.

230. To show the colour of iodine vapour. A few crystals are dropped into a large clean flask, which is warmed by a Bunsen. A smaller flask may be heated all over, and placed in the beam of the electric or lime light, and a few crystals of iodine introduced, when the violet colour of the vapour will be manifest on the screen.

231. To show the dichroism of iodine vapour. A small quantity of iodine, about ·25 gram, is put into a test tube. $15 \times 2\frac{1}{2}$ centimetres, and the tube drawn out and hermetically sealed. The tube is hung by a wire in a horizontal position in front of the lamp, and a narrow ray of light from a slit passed through it, an image of the slit being focussed upon the screen. On heating the tube the iodine vaporises, and the reddish violet colour will be seen. If the heat be continued, the colour loses its red tinge, and becomes a deep blue. Care must be taken to avoid having an excess of iodine in the tube, or before the blue becomes visible the quantity of iodine vapour present may render the tube practically opaque to the beam of light.

232. To show the action of iodine upon mercury. A small quantity of mercury and iodine are placed in a test tube and heated together. Combination takes place, and the iodide of mercury sublimes in the form of red and yellow crystals.

233. The dimorphism of mercuric iodide may be shown in the following way:—A quantity of dry precipitated mercuric iodide is placed upon a piece of filter paper and spread evenly over it by being rubbed with the finger. If the paper so coated

be gently heated over a small rose burner, the red compound is
rapidly transformed into the yellow crystalline variety. The
instability of the yellow or rhombic prismatic form at the ordinary temperature may be illustrated by drawing across the
paper a clean glass rod. Wherever the rod touches the paper
it leaves a streak of the red or quadratic crystals.

At low temperatures the yellow variety is the more stable,
and if a piece of blotting-paper smeared with the scarlet compound be wrapped round a large test tube in which a mixture of
solid carbon dioxide and ether is placed, the red crystals change
to the yellow form. By allowing the paper to project below the
tube the contrast is plainly visible. With the cold produced by
liquid oxygen the change is still more marked.

234. To show the action of iodine upon phosphorus. A few
crystals of iodine are brought in contact with a fragment of
phosphorus upon a block of wood. The phosphorus at once
inflames.

235. To show that phosphorus and iodine do not react if the
temperature is below that at which iodine gives off vapour.
A fragment of phosphorus and a few crystals of iodine in separate
test tubes are cooled by being placed in a freezing mixture of
ice and calcium chloride. The cooled substances may then be
mixed together and no combination takes place. If the tube
containing the two elements be removed from the freezing
mixture and allowed to become warm, combination takes place.

236. The action of iodine upon starch may be shown by
adding a few drops of a dilute solution of iodine, either in water
or potassium iodide, to a quantity of dilute starch emulsion contained in a large glass vessel.

The starch solution is prepared by mixing a few grams of
clean white starch with a small quantity of cold water, to the
consistency of a thin cream, and then rapidly pouring upon it a
large excess of boiling distilled water. The operation is best
performed in a beaker capable of holding forty or fifty times as

much water as was employed to mix with the starch in the cold.

237. The blue colour obtained in the above reaction is discharged on the addition of a few drops of chlorine water.

238. To show the effect of heat upon the blue compound When the solution is heated, the blue colour gradually disappears, but returns again with somewhat less intensity when the solution is cooled. For this experiment the starch solution should be largely diluted, and a small quantity of it filtered. To a few cubic centimetres of the clear solution a slight excess of a dilute solution of iodine in potassium iodide is added. If the blue so obtained be too dark, the mixture is diluted with water until the colour is of the requisite intensity. If a small quantity of this blue solution be gently heated in a test tube by being dipped in water at about 90°C., the colour will be rapidly discharged, and the solution become perfectly colourless. Upon dipping the test tube into cold water the blue colour is once more restored.

It will be found that if there be an excess of starch present in the solution, the colour that returns is far less intense than is the case when the iodine is in excess.

This experiment may be shown upon the screen, the hot water being contained in a glass cell with parallel sides.

239. To show the combustion of antimony in iodine vapour A small quantity of iodine is heated in a wide test tube, and powdered antimony is dusted into the vapour, when the metal will spontaneously ignite. The antimony should be contained in a small muslin bag, as described in Exp. 161.

HYDRIODIC ACID. HI.

240. Preparation by the action of water upon phosphorus iodide. $P + 5I + 4H_2O = H_3PO_4 + 5HI$. This may be done in a similar apparatus to that used for hydrobromic acid. A mixture of iodine and red phosphorus is placed in the dry flask, and water slowly dropped upon the mixture from the dropping funnel. The gas may be collected by displacement.

241. To prepare an aqueous solution of hydriodic acid see Hydrobromic Acid, Exp. 224.

242. To show the solubility of hydriodic acid in water see Hydrobromic Acid.

243. To show the action of chlorine upon hydriodic acid. Cylinders of the two gases are placed mouth to mouth and the covers removed. As the gases mix, iodine is liberated.

244. To show the effect of heat upon hydriodic acid. A flask filled with gaseous hydriodic acid is closed with a cork carrying two stout copper wires, which are joined at their lower ends by means of a loop of platinum wire soldered to each. An electric current is passed through the wire of sufficient strength to raise its temperature to dull redness, when the gas is decomposed and iodine vapour makes its appearance.

OXIDES AND OXYACIDS OF IODINE

245. Preparation of iodic acid by the action of nitric acid upon iodine. $3HNO_3 + I = H_2O + N_2O_3 + NO_2 + HIO_3$. Two or three grams of iodine are placed in a flask, and about 40 c.c. of fuming nitric acid (sp. gr. 1·5) added. On gently warming the mixture, oxides of nitrogen are rapidly evolved, and white crystals of iodic acid separate from the liquid.

(To purify the compound the mixture is evaporated to dryness, carefully heated to 200°C. to drive off all excess of nitric acid, dissolved in water and crystallised by evaporation.)

246. To show the reduction of iodic acid by sulphurous acid. $2HIO_3 + 5SO_2 + 4H_2O = 5H_2SO_4 + I_2$. Iodic acid has no action upon starch, but when brought into contact with sulphurous acid it is reduced, iodine being liberated, which may be seen by the formation of the blue compound with starch. This reaction appears to require time in order to start, but when the action once begins it proceeds with great rapidity. The phenomenon may be beautifully shown in the following way:—

10 grams of iodic acid are dissolved in 1 litre of distilled water. A small quantity of water is saturated with sulphur dioxide by bubbling the gas through it for a short time. 25 c.c. of this solution are diluted up to 1 litre with distilled water. These two standard solutions may be preserved. 50 c.c. of the standard iodic acid are added to 250 c.c. of distilled water in a large beaker, or cylinder, and a few c.c. of dilute starch emulsion added.

50 c.c. of the standard sulphurous acid are diluted with 250 c.c. water, and poured rapidly, to insure thorough mixing, into the solution of iodic acid. The mixture so obtained will remain colourless for thirty seconds, and then instantly, as with a flash, become blue throughout its entire mass. The interval of time which elapses before the reaction takes place can be lessened or increased by slight alterations in the dilution of the solutions.

FLUORINE

The preparation of fluorine, by the electrolysis of a solution of acid potassium fluoride in anhydrous hydrofluoric acid, is not very suitable for class demonstration, and should not be attempted without the greatest care being taken to prevent any of the vapour of the acid or the gaseous products of its decomposition from escaping into the air of the room.[1]

HYDROFLUORIC ACID. HF

247. The anhydrous acid may be prepared by distilling, in a platinum retort at a dull red heat, previously dried and fused acid fluoride of potassium, and condensing the liquid in a platinum condenser immersed in a freezing mixture of ice and salt. The intensely corrosive nature of this acid, and its extreme volatility, render it unadvisable to prepare it for class illustration. $HF,KF = KF + HF$.

248. To show the action of aqueous hydrofluoric acid upon glass. A sheet of glass is evenly coated with wax by brushing the melted wax over the warmed glass, and a design is then scratched upon the wax with the point of a fine steel knitting needle. If the glass be thinly coated it is quite easy to trace a design from a printed picture.

A sheet of blotting paper the size of the glass is laid upon it, and sufficient hydrofluoric acid poured over it to soak the paper. After about fifteen to twenty minutes the paper is removed, and the wax washed off by pouring hot water over the plate, when the design will be found to be etched into the glass. An image of the design may be projected upon the screen. The commercial acid supplied in gutta-percha bottles may be used for this experiment.

249. To etch glass with gaseous hydrofluoric acid. A glass plate prepared as above is placed over a shallow dish or tray made of sheet lead, or a platinum dish in which a quantity of finely-powdered fluorspar has been mixed with strong sulphuric acid. In about fifteen minutes the gas will have etched the design or writing.

[1] For a description of M. Moissan's experiments see *Ann. Chim. Phys.* [6] xii. 472.

SILICON FLUORIDE. SiF_4

250. To show the action of hydrofluoric acid upon silica or silicates. $SiO_2 + 4HF = SiF_4 + 2H_2O$. Finely-powdered fluor-spar mixed with about its own weight of sand is placed in a glass flask fitted with a safety funnel and delivery tube (fig. 74) and strong sulphuric acid added. On gently warming the mixture, silicon fluoride is evolved. If it be intended to carry on the operation for any length of time, either a thick glass bottle or one of stone-ware should be used and heated in a water-bath, as an ordinary glass flask will be corroded through in about an hour. The gas may be collected by displacement in a thoroughly dry cylinder.

251. To show the action of silicon fluoride upon combustibles. A lighted taper may be introduced into a cylinder of the gas, when the taper will be extinguished and the gas seen to be non-inflammable.

252. To show the action of silicon fluoride upon water. $3SiF_4 + 4H_2O = 2H_2SiF_6 + SiH_4O_4$. The gas is passed direct from the generator into water, when silicic acid will be seen to precipitate, and a solution of hydrofluosilicic acid is obtained. As the precipitated silicic acid would at once block up the delivery tube, it is necessary to prevent any water from gaining access to this tube. For this purpose mercury to a depth of two or three centimetres is first poured into the empty cylinder, and the dry delivery tube which dips beneath this mercury is then attached to the generating flask, and the cylinder then filled up with water. Each bubble of gas, as it rises through the mercury, and comes into contact with the water, is at once decomposed, and a little sack-like mass of silicic acid floats up to the surface of the liquid. If the heat be regulated so that the stream of gas is not too rapid, long stalactite-shaped masses of the silicic acid form upon the mercury, which often grow several centimetres long.

By arranging to perform this reaction in a narrow glass cell in front of the lamp, an image of these growing stalactites may be projected upon the screen.

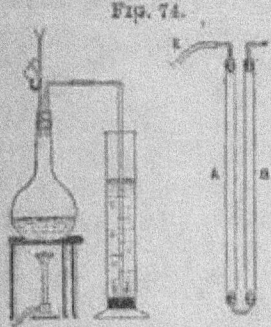

Fig. 74.

253. To show the deposition of silicic acid in a moist glass tube. Two long pieces of glass tube are connected at one end by a short bent glass tube, as shown in fig. 74, both tubes being dry. When the experiment is to be performed, tube B is disconnected, and the inside moistened by breathing through the tube; it is then replaced, and the flexible tube E attached to the generating flask. As the gas passes through the dry tube A it is not visible, but as soon as it enters the bottom of tube B a white film of silicic acid is deposited upon the glass, which steadily spreads up the tube as the heavy gas gradually rises to the top.

NITROGEN

PREPARATION.

254. By burning phosphorus in an inclosed volume of air. A fragment of phosphorus is placed in a small porcelain dish, which is floated in a trough of water; the phosphorus is inflamed and a bell-jar placed over it. As the flame dies out, the water rises in the bell-jar. It is convenient to employ a stoppered bell-jar, so that the gas may be more readily tested by means of a lighted taper. The bell must be depressed into the trough until the water outside is level with that inside, and then the stopper may be removed and a burning taper introduced. This experiment is not adapted for demonstrating the relative volumes of oxygen and nitrogen in the air, as the heat from the burning phosphorus at first expands the air in the bell-jar, and causes some of it to bubble out through the water, so that at the conclusion of the experiment the water will occupy considerably more than one-fifth of the volume of the vessel.

255. By a modification of the above experiment, in which the proportion of oxygen to nitrogen in the air is demonstrated. A long glass tube about $2\frac{1}{2}$ centimetres in diameter and 80 centimetres long, is closed at one end, and fitted with a cork at the other. The tube is graduated into five equal volumes by means of india-rubber rings slipped over it.

A dry piece of phosphorus is dropped into the tube, which is at once closed by the cork. The bottom of the tube is then

FIG. 75.

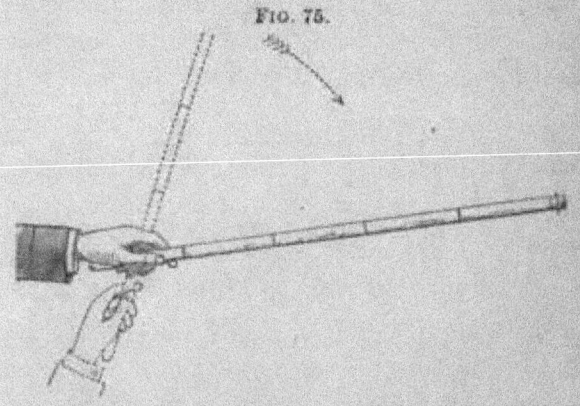

dipped into hot water, which melts and ignites the piece of phosphorus. The tube is now grasped near the bottom, and with a sudden jerk the melted phosphorus is flung along the entire length of the tube. In this way the whole of the oxygen is at once consumed. On allowing the tube to cool and opening the end under water the liquid will rise to the first ring.

In performing this experiment care must be taken to employ

sufficient phosphorus to flow the whole length of the tube, for should it not burn down the entire length some of the oxygen is liable to escape combination.

256. By absorbing the oxygen from an inclosed volume of air with cuprous chloride. A small quantity of cuprous chloride is poured into a stoppered cylinder, which is then closed and allowed to stand for a few hours. The absorption may be hastened by a gentle agitation of the cylinder, in order to moisten the sides of the vessel. The colourless cuprous chloride solution rapidly darkens as it absorbs the oxygen. The cylinder should be opened beneath the surface of water, which will enter to replace the absorbed gas, and the nitrogen may be tested by the usual method. (For Cuprous Chloride see Acetylene, No. 464.)

257. By absorbing oxygen from the air by heated copper. A piece of combustion-tube is packed with rolls of copper gauze, and heated to redness in a gas-furnace. A stream of air is passed over it, the air being contained in a gasholder, and

FIG. 76.

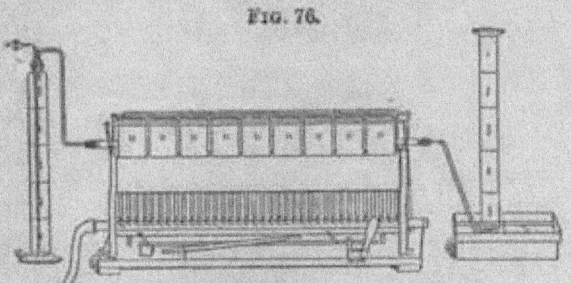

nitrogen is collected at the pneumatic trough. This experiment may be made roughly volumetric, by passing a measured volume of air through the tube, and measuring the gas collected. For this purpose two cylinders of equal size, and about 1 litre capacity, are graduated into five equal parts. One of them is fitted with a cork carrying two tubes; one descending to the bottom of the vessel, and the other passing just through the cork. The air from this cylinder is displaced and passed over the heated copper by allowing a slow stream of water to enter by the longer tube; the issuing gas is collected over water in the second cylinder. When the air has been entirely expelled, and the cylinder is completely filled with water, it will be seen that the gas collected in the receiving cylinder only occupies four of the five volumes.

NOTE. If, at the conclusion of the experiment, and while the tube is still hot, a stream of hydrogen or coal gas be passed through the tube for a short time, the oxide of copper which has been formed will be reduced, and the tube will thus be ready for use again.

258. By passing air and ammonia over red-hot copper. Air from a gasholder is made to bubble through strong ammonia in a Woulf's bottle, and thence through a tube containing heated copper gauze. The issuing nitrogen should be collected over water which has been acidulated with a small quantity of sulphuric acid. In this experiment the copper does not become oxidised, the oxygen being taken up by the hydrogen from the ammonia, so that the tube may be used as often as desired.

259. By the action of heat upon ammonium nitrite. $NH_4NO_2 = 2H_2O + N_2$ or $NH_4Cl + NaNO_2 = NaCl + 2H_2O + N_2$. A small quantity of ammonium chloride (about 10 grams is a convenient weight) is placed in a small flask fitted with a cork and delivery tube, and about one and a half times its weight of sodium nitrite added; the flask is about one fourth filled with water, and the mixture gently warmed. When the reaction sets in, the lamp may be removed, and the action allowed to continue without further application of heat. Should the effervescence become too violent, a small dish of cold water may be brought up under the flask, which at once checks the action.

260. By heating a mixture of ammonium chloride and potassium bichromate. $(NH_4)_2Cr_2O_7 = Cr_2O_3 + 4H_2O + N_2$. A mixture of the two salts, in the proportion of one part of ammonium chloride to three of potassium bichromate, is introduced into a short piece of combustion-tube, closed at one end, and fitted at the other with a cork and delivery tube. The tube should be held in a horizontal position and heated by means of a Bunsen, and the gas collected over the pneumatic trough.

261. By the action of chlorine upon ammonia. $4NH_3 + 3Cl = 3NH_4Cl + N$. Into one neck of a Woulf's bottle is fitted a straight piece of glass tube, as wide as the neck will conveniently allow; the tube passes nearly to the bottom of the bottle, and may extend about the same distance above; into the top of this tube

FIG. 77.

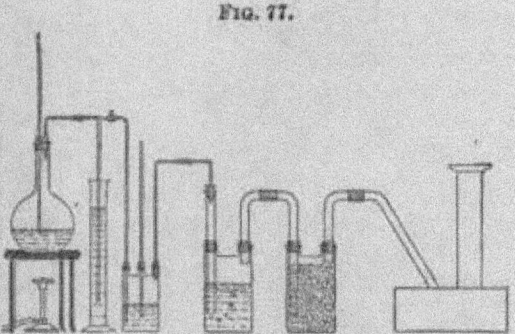

is fitted a cork with a small tube bent at right angles, by means of which connection is made with the chlorine apparatus. Into the second neck of the bottle an exit tube is fixed, also made of wide glass tube. A second Woulf's bottle, similarly fitted with wide glass tubes and filled with broken glass moistened with water, is necessary to wash the gas from the ammonium chloride which is produced in the reaction. The object of fitting the apparatus with tubes of large bore is to prevent stoppage from the deposition of ammonium chloride.

The first Woulf's bottle is about two-thirds filled with strong ammonia solution (sp. gr. ·880), and a rapid stream of chlorine, which should be first washed through water, is passed through it. As each bubble of chlorine enters into the ammonia the combination is attended by a feeble yellowish flame, which is best seen in a darkened room, and a rapid stream of nitrogen is evolved. This reaction must not be allowed to proceed until the ammonia is all decomposed, as at that point the dangerous compound, nitrogen chloride, begins to form ; for this reason it is well to employ a somewhat large quantity of the ammonia solution.

262. To show the behaviour of nitrogen towards combustibles. A lighted taper may be introduced into a vessel containing the gas, when the flame is extinguished, and the gas is not inflamed.

263. To show the combination of nitrogen and oxygen by the action of electric sparks. A three-necked glass globe, of about 200 c.c. capacity, is connected to a small condensing syringe capable of producing a pressure of a few atmospheres. The remaining two necks are fitted with corks carrying glass tubes to which platinum wires are sealed. A stream of electric sparks is passed into the inclosed air, and the pressure gradually increased by means of the pump. In the course of a short time the air within the globe will be seen to be distinctly coloured with the oxides of nitrogen, which will be the more visible if the vessel be placed upon a white surface. It is necessary to wire the corks into the several necks.

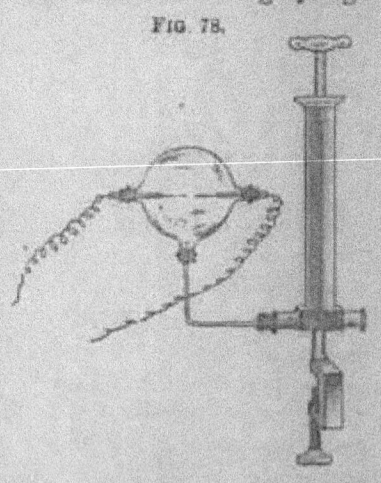

FIG. 78.

264. This experiment is more strikingly shown by employing a glass cylinder fitted with plain glass ends, and passing a ray of light through its length, the light being received upon the

screen. For this purpose a cylinder about 6 or 7 centimetres wide and 30 centimetres long, has two small holes drilled into it near the middle, and opposite to each other; into these holes, which are readily cut by means of the point of a triangular file with a little fine emery and turpentine, two platinum wires are inserted. This may be done by passing each wire through a cork,

FIG. 79.

one end of which is slightly hollowed to fit the curved side of the cylinder, to which it is attached with a little cement. Very convenient ends or caps for the cylinder may be made of the covers of ordinary round glass-covered specimen boxes.

The cylinder is placed in front of the lamp and a parallel beam of light passed through it. By means of a focussing lens, an image of the end farthest from the lamp is projected upon the screen, and a stream of electric sparks passed between the two wires. The white colour of the disc upon the screen will be seen to rapidly become yellow and then brownish in colour as the passage of the sparks is continued. While the light is still passing, one of the caps may be removed from the cylinder, and the gaseous contents rapidly swept out by means of a stream of air from a foot-bellows, when the original white colour of the light will be seen in striking contrast.

265. To show the combination of nitrogen with oxygen by the action of the oxy-hydrogen flame. A jet of hydrogen is burnt in air containing an excess of oxygen, and the gaseous products tested with starch and potassium iodide. For this purpose a common fish-tail burner is fixed upon the end of a piece of iron or brass pipe inserted through a cork. Over the cork is fitted a wider piece of glass tube, which forms a jacket to the metal tube, and ends at the base of the burner; through the same cork is fitted a short piece of small glass tube, through which oxygen can be passed. A small flame of

hydrogen is burnt at the jet, and a gentle stream of oxygen allowed to pass in until the flame shows round its base the bluish colour characteristic of its combustion in oxygen. A clean dry cylinder is then inverted over the flame for a few seconds, to allow of the collection of the products of combustion, and a strip of moistened 'starch paper' introduced, when the paper will be instantly rendered blue. The quantity of the oxides of nitrogen produced is sufficiently abundant to be readily perceived by the smell.

FIG. 80.

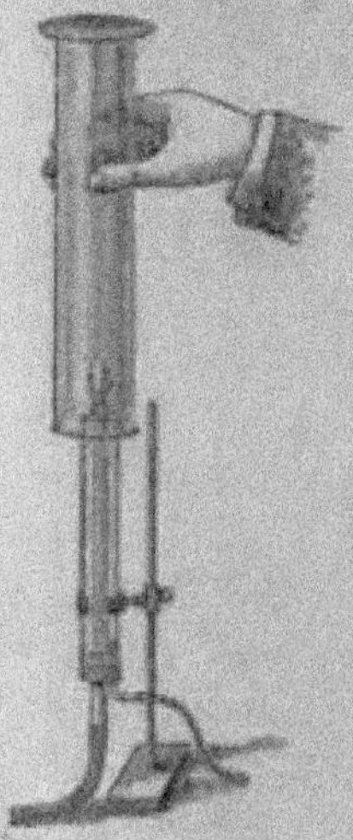

If the products from this combustion be collected as already described (Hydrogen, Exp. 24), the acidity of the water may be demonstrated, as well as its action upon potassium iodide and starch.

266. To show the combination of nitrogen with oxygen by the action of burning magnesium. A strip of magnesium ribbon from six to ten centimetres long is folded up and attached to a deflagrating spoon; the metal is ignited and lowered into a clean dry cylinder, and there allowed to burn itself out. On introducing a moistened starch paper into the vessel the blue colour will at once be produced.

Should it be desired to test the contents of the cylinder by means of a solution containing starch and potassium iodide in place of the paper, a few drops of very dilute sulphuric acid must be added.

267. To show the combination of magnesium with nitrogen. When powdered magnesium is strongly heated in nitrogen, the metal combines with the gas forming magnesium nitride. N_2Mg_3. A piece of combustion tube about 18 c.m. long is drawn out at one end, and the narrow portion bent at right angles; to this is attached a piece of small tube about 70 c.m. long, which dips into a small beaker containing coloured water.

A small quantity of magnesium filings is placed in the combustion tube, and a brisk stream of nitrogen passed through the

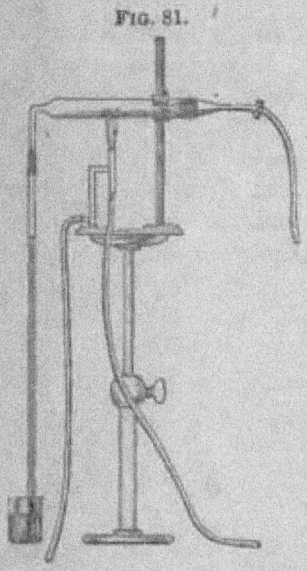

FIG. 81.

tube to sweep out all the air. The tube containing the metal is then gently heated with a Bunsen, while a slow stream of nitrogen is still passing. When the temperature of the metal approaches to redness, the passage of the nitrogen is stopped, and the Bunsen replaced by a Herapath blowpipe, which is being gently fed with oxygen. This quickly raises the temperature of the magnesium to the required degree, and the absorption of the nitrogen begins to take place as indicated by the rapid rise of the water in the tube. As the liquid reaches the top of the tube a little more nitrogen may be allowed to enter, and in this way the absorption of the gas may be continued for some time.

268. The decomposition of the magnesium nitride by hydrogen, and the formation of ammonia, may be demonstrated by passing a gentle stream of hydrogen through the tube while the nitride is still warm. The tube conveying the nitrogen is disconnected from the reservoir containing that gas, and connected to a hydrogen bottle, and a slow current allowed to pass over the compound. The long tube dipping into the water is removed, and the issuing gas allowed to impinge upon turmeric paper, when abundant evidence of the presence of ammonia will be seen.

269. The nitride may also be decomposed by water, with the formation of ammonia. $N_2Mg_3 + 3H_2O = 3MgO + 2NH_3$. For this purpose the nitride may be transferred to a test tube

and warmed with a little water, the ammonia being shown by
means of turmeric paper.

270. **To show the combustion of lithium in nitrogen.** When
the metal lithium is heated to dull redness in a stream of
nitrogen it undergoes a feeble combustion, combining with the
gas to form lithium nitride NLi_3.

A small quantity of lithium, contained in a shallow boat
made of thin sheet iron, is placed in a piece of combustion
tube, and a stream of nitrogen passed over it. On heating the
tube by means of a Bunsen the lithium combines with the
nitrogen, the combination being attended with a feeble glow.
In a few minutes the whole of the metal will have become
burnt.

As in the case of the magnesium above described, if a
stream of hydrogen be passed through the tube while it is
still being heated, torrents of ammonia will be evolved, which
may be rendered evident by means of turmeric or red litmus.

ARGON

271. The actual isolation of argon is a process which
occupies too long a time to allow of it being conveniently employed as a lecture experiment. The principle involved, however, in the two methods for separating atmospheric nitrogen
from argon may be illustrated. The removal of nitrogen by
means of magnesium may be exemplified by means of Experiment No. 267. But a much more rapid absorption of nitrogen
is obtained by substituting for magnesium a mixture of 1 gram
powdered magnesium, 5 grams dry powdered lime (calcium
oxide) and ·25 gram sodium, cut in small pieces. The method of
sparking a mixture of air and oxygen, and the absorption of the
nitroxygen compound by potash, may be shown in the following
way. A small tudulated glass basin (which may conveniently

be made by cutting down an ordinary narrow-mouth bottle) is
fitted with a cork through which pass two glass tubes. Through
these tubes pass two platinum wires, which are fused through
the upper ends of the tubes. The ends of these wires are melted
in an oxyhydrogen flame so as to obtain a little knob of platinum
on the end of each (this prevents the wires becoming so quickly
heated when the sparks are being passed). A solution of caustic
potash is placed in the dish, and a narrow
test tube is inverted over the wires. In
order to bring the liquid up to the desired
height inside the tube, a piece of fine
rubber tube should be previously pushed
up and the air sucked out until the
potash is within about an inch of the top.
The rubber tube is then withdrawn and a
little oxygen bubbled up, about equal in
volume to the air. The height of the
liquid is now marked by an indiarubber
band on the test tube. The apparatus
(fig. 81ᴀ) is then placed in front of the
lantern and an image of the tube thrown
upon the screen. The quantity of gas
should be so arranged that the length of
the tube from the band to the top is well
within the field. A stream of strong
sparks (using a Leyden jar with the coil)
is passed between the platinum wires,
and although the first effect is the ex-
pansion of the gas, due to heating, the liquid will gradually
ascend in the tube as the nitrogen and oxygen unite and the
oxide is absorbed by the potash.

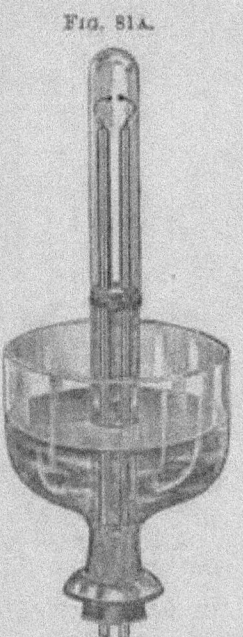

Fɪɢ. 81ᴀ.

AMMONIA. NH₃

272. **To show the synthetical formation of ammonia.** For this purpose the ozone apparatus already described (Ozone, Exp. 61) may be used. A mixture of three volumes of hydrogen and one volume of nitrogen is made in a small gasholder, and the gas, after bubbling through a small wash-bottle containing

FIG. 82.

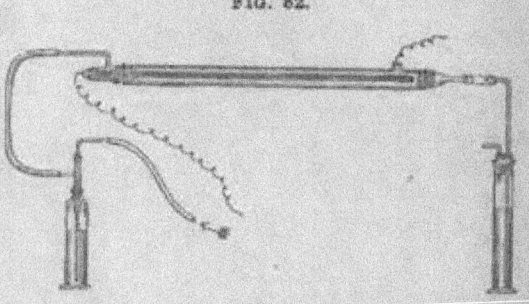

dilute sulphuric acid, is slowly passed through the apparatus and exposed to the electric discharge. The issuing gas is made to bubble through Nessler's solution contained in a narrow cylinder bottle standing upon a white surface. When the gas has been passing for some little time, the Nessler's solution will gradually become tinted of a yellowish-brown colour, which may be contrasted with a similar bottle containing some of the unacted-upon solution. The action being slow, the passage of the gas should be allowed to continue for half an hour or longer.

273. **Preparation of ammonia gas from the solution.** Strong ammonia solution (sp. gr. ·880) is placed in a flask fitted with a cork carrying a mercury safety funnel and a delivery tube; the gas is passed through a Woulf's bottle filled with fragments of

quicklime, and may be collected over mercury, or by upward displacement.

274. **Preparation of ammonia by the action of lime or soda lime upon nitrogenous organic matter.** Horn shavings, gelatin, or glue is mixed with soda lime in a dry flask or test tube and gently heated. Ammonia in abundance is evolved, which may be seen by allowing the escaping gases to impinge upon moist turmeric paper.

275. **Preparation of ammonia from ammonium chloride.** $2NH_4Cl + CaH_2O_2 = CaCl_2 + 2H_2O + 2NH_3$. Ammonium chloride mixed with about half its weight of lime is gently heated in a flask. The gas may be collected by displacement, or a solution in water may be made; in the latter case the gas should be first passed through a small empty three-necked Woulf's bottle fitted with a mercury safety funnel, and then into two others in series, which are half filled with water.

276. **To show the lightness of ammonia.** A cylinder of the gas may be poured upwards into an inverted beaker, suspended from the beam of a balance. (See Hydrogen, Experiment 13.)

Or a collodion balloon may be filled with the gas, and allowed to ascend.

277. The behaviour of ammonia towards combustibles may be shown by introducing a lighted taper up into an inverted cylinder of the gas. If the taper be cautiously introduced, the combustion of the ammonia round the flame of the taper, with its characteristic yellowish flame, will be seen. When thrust into the gas, the taper is extinguished.

278. **To show the combustion of ammonia in oxygen.** A short piece of wide glass tube, about 12 centimetres long and 4 centimetres diameter, is fitted at one end with a cork, carrying two tubes, bent at right angles. One of these passes just through the cork, while the other extends to the end of the outer glass cylinder, the latter tube being in the centre of the cork. A

K

small flask containing strong ammonia solution, and fitted with a cork and delivery tube, is connected to the central tube, and a moderate stream of ammonia made to issue by applying a gentle heat to the flask. On applying a lighted taper to the issuing ammonia, its curious behaviour as regards combustibility may be seen. If a gentle stream of oxygen be now caused to enter the apparatus by the short tube, the ammonia readily ignites at the jet, and continues to burn with a yellowish-brown coloured flame. In order to distribute the oxygen, and cause it to surround the central tube, a little plug of cotton wool may be packed loosely into the apparatus. As soon as the oxygen supply is cut

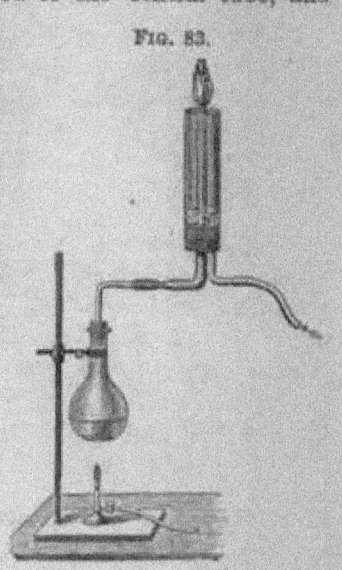

Fig. 83.

off, the flame of the burning ammonia languishes and dies out. In order to prevent the flame from being tinged yellow by the hot glass jet, and the true colour of the burning ammonia being thereby masked, it is well to insert into the jet a small roll of platinum foil, which should project slightly beyond the glass.

279. To show the combustion of ammonia in chlorine. A little strong ammonia solution is heated in a flask, to which is connected by a flexible tube a bent-glass jet. On lowering the jet into a jar of chlorine the gaseous ammonia spontaneously ignites, and continues burning with a yellowish flame.

280. To show the decomposition of ammonia by heat. A stream of the gas is passed through a heated tube, and the mixed nitrogen and hydrogen collected over water acidulated with sulphuric acid.

If a furnace be used in which to heat the tube, a piece of narrow iron gas pipe answers very well, but if the experiment is to be made upon a small scale, a short platinum tube (see Experiment 7), heated by a Bunsen lamp, will give an excellent result. A piece of hard glass tube may be substituted, but with a smaller effect.

281. To decompose ammonia by electric sparks, and show the volume composition. Into the closed limb of a U-shaped eudiometer is introduced a quantity of ammonia gas. (For the method of filling the tube, see Water, Experiment 74.)

The eudiometer is divided by bands into two parts, one band marking the volume of the included gas when the mercury is at the same level in both limbs, the other being placed at twice the distance from the top. On passing a stream of electric sparks, the gas at once expands, partly owing to the heat and partly to its decomposition. The sparks may be continued for a few minutes, the gas being allowed to remain under the increased pressure caused by the rising of the mercury in the open limb, as this accelerates its decomposition. When the action is finished, and the tube allowed to become cold, it will be seen, on levelling the mercury by running the excess away by the side tube, that the volume of the gas has doubled. The gas may be decanted into the open limb of the eudiometer by first filling with mercury, and then closing the end with the thumb, and inverting the apparatus. When all the gas has been passed into the open limb, a lighted taper may be applied to show that hydrogen is present in sufficiently large proportion to enable the gas to burn.

282. To show the volumetric composition of ammonia. A straight glass tube, about 50 centimetres long, is fitted at one end with a cork, carrying a stopcock and two stout copper wires. To the ends of these wires is fastened a platinum wire wound into the form of a conical spiral, which can be heated by the passage of an electric current. This platinum spiral, which is practically a little basket or cage (fig. 84), is filled up with frag-

ments of copper oxide. The oxide made from wire answers very well. The tube is graduated into four equal parts by means of india-rubber rings, the bottom one being about 10 centimetres from the end of the tube. The platinum cage must not reach further into the tube than the top division. The tube is stood in a trough of water, dipping one or two centimetres beneath the surface, and a stream of the mixed gases, obtained by passing ammonia through a red-hot tube, is passed in by the stopcock at the top until the air in the tube has been entirely swept out. The gases should be passed through a Woulf's bottle, containing dilute sulphuric acid, to wash them free from any undecomposed ammonia before being delivered into the eudiometer. When the tube is filled, the stopcock is closed, and the tube lowered in the trough until the water outside is level with the lowest band on the tube, and the pressure restored by momentarily opening the cock. The platinum cage is then heated by means of an electric current, and as the copper oxide becomes warm the hydrogen is absorbed. When the action is completed and the tube allowed to cool, the water will have risen to the top band, three volumes of hydrogen having been absorbed, and one of nitrogen remaining.

FIG. 84.

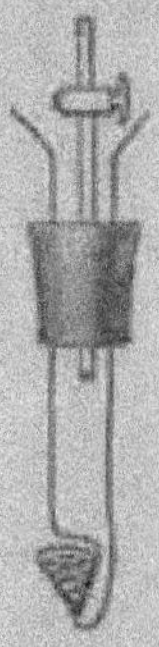

283. To show the volumetric composition by means of chlorine. $2NH_3 + 3Cl_2 = 6HCl + N_2$. A glass tube about one metre long, closed at one end, and divided into three equal volumes by means of india-rubber rings, is fitted with a cork carrying a small stoppered dropping funnel. The tube is filled with chlorine by collection over a saturated solution of salt in water, and the cork inserted. A few cubic centimetres of strong ammonia solution are poured in the funnel, and allowed slowly to enter into the tube drop by drop. As the first two or three drops fall into the tube, it will be seen that the combination is attended with a feeble flash of light. When the decomposition is complete, the funnel is filled up with water acidulated with

sulphuric acid, and a glass tube bent twice at right angles is fitted by means of a small piece of caoutchouc tube into the neck of the funnel, the other end of the tube dipping into a beaker of water. On opening the stopcock, water will be drawn in, filling two thirds of the tube. The remaining one volume may be shown to be nitrogen, by introducing a lighted taper. This one volume of nitrogen was therefore in combination with the three volumes of hydrogen which have now combined with the three volumes of chlorine originally contained in the tube. The resulting hydrochloric acid, in the presence of the excess of ammonia, forms the solid ammonium chloride, and therefore ceases to have any volumetric significance.

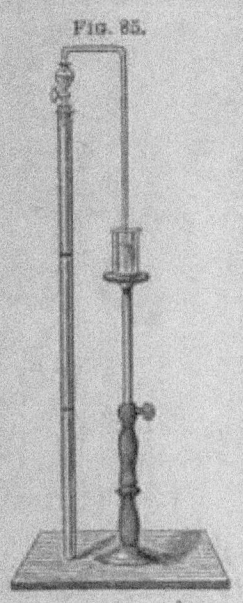

Fig. 85.

284. To show the volume composition of ammonia by electrolysis. A solution consisting of one volume of strong ammonia (sp. gr. ·880), and ten of a saturated solution of salt in water, is subjected to electrolysis in any convenient electrolysis apparatus (see Water, Experiment 78) by a current of 5 to 10 Grove's cells. The gases will be evolved in the proportion of three to one.

285. To burn potassium in ammonia. A gentle stream of ammonia is passed over a fragment of potassium contained in a combustion bulb. On applying a gentle heat by means of a Bunsen lamp, the potassium bursts into flame, and the hydrogen may be inflamed as it issues from the tube.

286. To show the solubility of ammonia in water. (See Hydrochloric Acid, Experiment 193.)

287. To show cold produced by the evaporation of ammonia solution. A small wide-necked flask containing a quantity of

strong ammonia solution is placed upon a wooden block, upon which a few drops of water have been poured, and a rapid stream of air from a foot bellows blown through it. In a few moments the flask will be frozen to the block.

FIG. 86.

288. To liquefy ammonia by cold and pressure. On a small scale this may be shown by Faraday's method. (See Liquefaction of Gases, Experiment 644.)

On a larger scale liquid ammonia may be obtained by passing the gas, evolved from the strong aqueous solution, through a cold U tube, under the pressure of an extra atmosphere.

A quantity of strong ammonia is placed in a strong round-bottomed flask, fitted with a cork, through which pass two tubes. One of these is bent at right angles, and is connected with a U tube, having a stopcock on each limb. The other is bent twice at right angles, and is of such a length that it can reach to the bottom of a wide tube about a metre long, which is nearly filled with mercury (fig. 87).

FIG. 87.

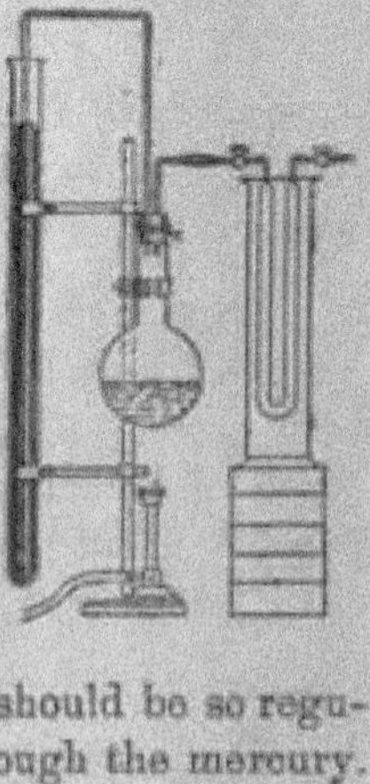

The cork is securely wired into the flask and a gentle heat applied, the air being allowed to escape through the U tube, which is surrounded with a freezing mixture of ice and calcium chloride. The cock upon the limb of the U tube furthest from the flask is then closed, and as the internal pressure gradually increases bubbles of gas will escape through the column of mercury. If the temperature of the freezing mixture be below −20° C. and the column of mercury as much as 760 mm., liquefaction of the gas rapidly takes place in the U tube. The heat should be so regulated that only a very little gas escapes through the mercury. When sufficient liquid has been collected the lamp is removed, the second cock closed, and the mercury tube gently lowered in order gradually to relieve the internal pressure within the flask. The U tube may then be disconnected and removed from the freezing mixture.

289. To liquefy ammonia by cold. A stream of ammonia gas, obtained from the strong solution, is passed through a gas-condensing tube (see Liquefaction of Gases, Exp. 649), which is immersed in a mixture of solid carbon dioxide and ether. A convenient vessel for the purpose is a boiling tube, just large enough to admit the condensing tube, round which is wrapped a piece of flannel or green baize. The tube is about half filled with ether, and a quantity of solid carbon dioxide introduced. A rapid stream of ammonia gas may be passed through, as at the low temperature of the bath condensation is very complete.

290. To show the solubility of sodium or potassium in liquid ammonia. The liquid ammonia for this experiment may be prepared as above, but as it is necessary for the liquid to be as dry as possible, the ammonia should be passed through a Woulf's bottle containing quicklime before it is condensed. The tube containing the liquid is removed from the bath, and a fragment of potassium introduced; the metal is instantly dissolved, forming a deep indigo blue solution. The solution cannot be preserved, as in a short time the ammonia is acted upon by the potassium.

291. To show the oxidation of ammonia by means of heated platinum. A spiral of platinum wire, which may be conveniently hung from a piece of asbestos cardboard, is heated and suspended in a jar containing a few cubic centimetres of strong ammonia. The wire at once begins to glow, and white fumes of ammonium nitrite are formed. On blowing oxygen very gently into the top of the jar, by introducing a delivery tube through a hole cut in the cardboard, ruddy fumes begin to appear, and on increasing the supply of oxygen the platinum becomes hotter, hydrogen is evolved, and a small deflagration results.

292. To show the combination of ammonia with hydrochloric acid. Two cylinders, one filled with gaseous hydrochloric acid and the other with ammonia, may be placed mouth to mouth, when combination at once takes place with the formation of solid ammonium chloride.

The formation of ammonium chloride may also be shown by passing gaseous hydrochloric acid into a strong solution of ammonia. The vessel containing the ammonia should be placed in another vessel of cold water, and the ammonium chloride crystallises out as it is formed.

293. To show the volumetric combination of ammonia with hydrochloric acid. Exactly equal volumes of the two gases are separately collected in two small graduated cylinders over mercury. The contents of one cylinder are then decanted up into the other, care being taken that no gas escapes.   When the entire quantity has been transferred, the second cylinder is carefully removed from the trough by means of a glass plate, when it will be seen to be completely full of mercury, the whole of the gas having disappeared.

294. To show the heat evolved by the union of ammonia with hydrochloric acid gas.   A large flask, having two side tubulures,

FIG. 88.

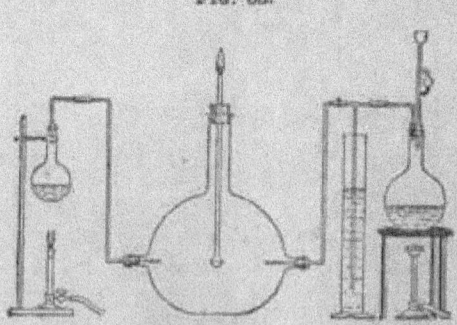

is connected by one to a hydrochloric acid apparatus, and by the other to an ammonia-generating apparatus. The first of these pieces of apparatus should be fitted with a three-way tube, so that the stream of hydrochloric gas can be stopped or regulated at will. Into the mouth of the flask a loosely-fitting cork is inserted, carrying a long tube with a bulb blown upon the end, and in which a small quantity of ether is placed. The cork must not completely close the flask; it should therefore be either bored with a second hole, or have a channel cut upon its side. A stream of hydrochloric acid and one of ammonia are allowed to enter the flask, and as they combine to form the white cloud of ammonium chloride sufficient heat will be generated to evaporate the ether contained in the little bulb, and the vapour may be inflamed as it issues from the end of the tube.

The experiment can be performed in an ordinary flask, instead of the more expensive vessel described. In this case the tubes delivering the two gases must pass through the cork which also carries the bulb. One tube should extend nearly to the bottom of the flask, while the other, which conveys the hydrochloric acid, should reach only about one-third of the way down.

295. To prepare the so-called ammonium amalgam. An amalgam of mercury and sodium is first prepared by gently pressing a few fragments of sodium into a mortar containing some mercury, by means of the pestle. The combination of the two metals takes place with some energy, and unless the sodium is buried beneath the surface of the mercury it is liable to ignite and be thrown out of the mortar. The liquid amalgam is then poured into a large dish half filled with a moderately strong solution of ammonium chloride, which may be slightly warm. The metallic mass at once swells outs and floats upon the liquid as a sponge, at the same time evolving large quantities of ammonia gas, which may be shown by holding a turmeric paper over the dish.

296. To show that hydrogen is evolved from the 'amalgam,' a lump of it is placed in a cylinder full of water, and the cylinder

then inverted in a dish or trough, the amalgam being prevented from escaping by means of a ground-glass plate; bubbles of gas will be seen to rise from the metal as it gradually shrinks in bulk, and by the application of a taper it may be demonstrated that the gas is hydrogen.

297. **To show the effect of pressure upon the 'amalgam.'** A small quantity of sodium amalgam is poured into a solution of ammonium chloride contained in a glass cell, which is connected to a small compression pump, an image of the cell being thrown upon the screen (see Carbon, Exp. 397). The amalgam of sodium should be of such a composition that the 'ammonium amalgam' which is formed will only just float up; this is arranged by trial. About 2 c.c. of the sodium amalgam are added to a sample of the ammonium chloride solution, and if the amalgam that is formed is too buoyant a small quantity of the sodium amalgam is diluted with mercury until a suitable strength is obtained. About 2 c.c. of the amalgam is sufficient for the experiment. It is thrown into the cell, and the top quickly screwed down. The amalgam then floats up to the surface of the liquid, and on subjecting it to a little pressure, by one or two strokes of the condensing pump, it will be seen to shrink in bulk, and sink in the liquid; on releasing the pressure the mass instantly re-expands, and again floats up to the surface.

HYDRAZOIC ACID

Azoimide. N_3H.

298. **Preparation by inorganic reactions.** This interesting substance may readily be obtained, in quantity sufficient to illustrate some of its most remarkable properties, by heating

sodium in dry ammonia gas, and then acting on the sodamide so produced, by nitrous oxide. These two reactions yield the sodium salt, from which the acid itself can readily be obtained.

1. $NH_3 + Na = NaNH_2 + H$

2. $\begin{cases} (a)\ NaNH_2 + N_2O = NaN_3 + H_2O \\ (\beta)\ NaNH_2 + H_2O = NaHO + NH_3 \end{cases}$

A small quantity of sodium (1 to 1½ grams) is placed in a silver boat, and introduced into a short piece of combustion tube closed

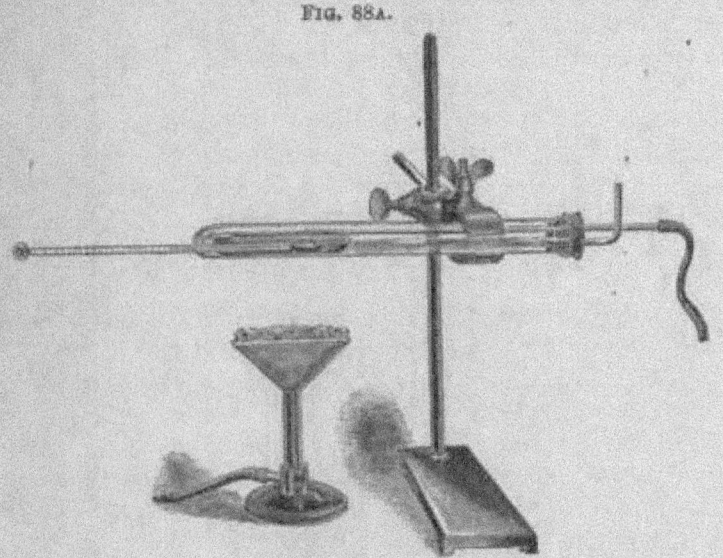

FIG. 88A.

at one end, and through which a stream of dry ammonia can be passed (fig. 88A). The ammonia is dried by being passed through two large U tubes containing pieces of quicklime, one smaller one filled with caustic potash, and finally through a tube containing pieces of metallic sodium.

As soon as the air is all expelled from the apparatus the tube containing the boat is gently heated with a Bunsen lamp, when the sodium melts and presents a perfectly bright silvery surface. In a few moments drops of a greenish-brown liquid begin to collect on the surface of the metal—this is the sodamide in a state of fusion, and if the stream of ammonia is not too rapid, the hydrogen which passes out of the apparatus may be inflamed at the exit tube. In a short time the whole of the sodium will be converted into the amide. When this point is reached the lamp is removed, and the drying-tubes are disconnected.

The next stage of the process consists in heating the sodamide to 200°C. in a stream of nitrous oxide.

The nitrous oxide (which is most conveniently derived from a cylinder of the liquefied gas) is bubbled through sulphuric acid, and then passed through the tube containing the sodamide. The tube may be heated by means of a flat-flamed Bunsen placed at some little distance below it, and the temperature may be controlled either by lashing a thermometer to the tube or holding it alongside by means of a clamp. The temperature should be kept about 200°, and should not be allowed to rise above 250°. The reaction (which takes about an hour for its completion) is accompanied by the evolution of ammonia, and is complete when this gas is no longer given off.

The tube is then allowed to cool, and the contents dissolved out with a small quantity of water. The solution is acidified with dilute sulphuric acid and the mixture distilled in a small Wurtz flask connected to a condenser, until the distillate is no longer acid. The distillate consists of a solution of hydrazoic acid, which in a cool dark place can be preserved.

299. To show the combination of hydrazoic acid with ammonia. $NH_3 + HN_3 = NH_4N_3$. If a portion of the acidified liquid prepared for distillation in the above experiment be heated in a test tube, and a glass rod moistened with ammonia be held near the mouth of the tube, white fumes of the ammonium salt, NH_4N_3 or N_4H_4, will be obtained.

300. To show the explosive nature of the salts of hydrazoic

acid. The most convenient compound is the silver salt. Silver nitrate solution is added to a small portion of the solution of the acid, obtained as described above, when a white curdy precipitate of silver hydrazoate (AgN_3) is at once thrown down. The precipitate, after being washed upon a filter, is collected together with a little wooden splint or a bone spatula, and gently pressed between blotting paper to remove the excess of water. It should then be placed in small quantities on separate pieces of filter paper to dry. On gently heating one of these over a small flame, the compound explodes with an extremely sharp report. A quantity of the salt equal in bulk to a split pea is quite enough to produce a very considerable explosion.

NITROUS CHLORIDE. NCl_3?

301. For lecture demonstration this substance should only be prepared on a very small scale, and the compound exploded as fast as it is formed. This may be safely achieved in the following way.

A glass cylinder about 15 centimetres long and 6 centimetres wide is closed at one end by means of a piece of bladder tightly stretched over it. A convenient apparatus is made by cutting the foot off an ordinary gas-collecting cylinder, and tying the bladder over the lipped end. The cylinder is stood upon three pieces of cork in a small glass dish nearly filled with a saturated solution of ammonium chloride; the cylinder is then half filled with more of the same solution. (Fig. 89.)

A platinum electrode, fused into a glass tube which is filled up with mercury, is placed within the cylinder, and a second electrode is placed in the dish below the cylinder. After the electrode has been put into the cylinder, a shallow layer of turpentine is floated on the surface of the aqueous solution. It is necessary that the electrode does not come in contact

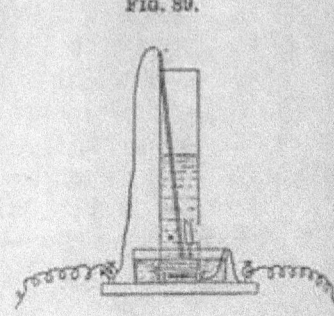

FIG. 89.

with the turpentine, so that it must not be introduced *after* the turpentine has been poured in. This electrode is then connected with the positive wire from a battery consisting of 10 Grove's cells, the negative wire being connected to the electrode in the dish. The chlorine which is evolved at the positive plate reacts upon the ammonium chloride, producing nitrous chloride, which as fast as it is formed is floated up by the bubbles of gas, and, coming in contact with the layer of turpentine, is decomposed with a slight crackling noise. Occasionally the decomposition is attended with a small detonation, which often inflames the turpentine.

Care must be taken that no mistake is made in connecting the electrodes with the battery, for, should the outside plate by accident be made the positive electrode, dangerously large quantities of the compound might accumulate in the apparatus before the mistake was discovered.

NITROGEN IODIDE
NHI$_2$.

302. This substance is best prepared by digesting powdered iodine for a few minutes in strong ammonia (sp.gr. ·880), the mixture being stirred with a stick of wood, or a small wooden spatula.

$$3NH_3 + 2I_2 = NHI_2 + 2NH_4I.$$

After a short time the mixture is filtered, and small quantities of the solid substance placed upon filter paper, and left to dry spontaneously. So long as the compound is moist it is not explosive, but when perfectly dry it explodes with very slight cause; for this reason three or four little quantities of the substance should be put on separate blocks of wood, and placed at some distance from each other, so that should one of them be exploded by a fly, or a falling particle, the others may not be affected by the shock. When quite dry the extreme explosibility of the compound may be shown by touching one of the small heaps with a feather tied to a stick. If a few minute particles of the substance, in the damp state, be thrown upon a sheet of paper, and then allowed to dry, a sudden puff of breath upon them, strong enough to blow them along the paper, will make them all explode.

303. To show the action of light upon nitrogen iodide. A small quantity of the compound is placed in a test tube, the end of which has been somewhat enlarged by being slightly blown out in the form of a bulb, and the tube filled up with dilute ammonia. When the powder has entirely settled to the bottom, the tube is placed in a broad glass cell containing water, and an image of the tube projected upon the screen by means of the electric light. As the light continues to shine upon the tube, bubbles of gas will be seen to be disengaged from the solid at the bottom of the tube. The object of immersing the tube in a vessel of water is twofold—viz., to prevent the heating of the liquid by the lamp, and also to enable a better image of the tube to be thrown upon the screen.

OXIDES OF NITROGEN

NITROUS OXIDE. N_2O.

304. Preparation from ammonium nitrate. $NH_4NO_3 = 2H_2O + N_2O$. A quantity of the crystallised salt is placed in a small flask fitted with a delivery tube. On the application

of a gentle heat, the salt melts and rapidly breaks up into water and nitrous oxide. The gas may be collected over warm water.

305. To show the behaviour of nitrous oxide towards ordinary combustibles. A lighted taper thrust into a cylinder of the gas burns with increased brilliancy.

306. A piece of phosphorus contained in a deflagrating spoon, when ignited and introduced into nitrous oxide, burns much as it does in oxygen.

307. A piece of sulphur (see Oxygen, Exp. 46) when just ignited may be extinguished by being thrust into a cylinder of nitrous oxide. If the sulphur be withdrawn, and allowed to burn freely in the air before being placed in the gas, it will then continue its combustion with considerable energy.

308. To distinguish between oxygen and nitrous oxide. Two tall cylinders are each half filled with nitric oxide over water; into one of them a cylinder of nitrous oxide is decanted, and to the second a similar cylinder of oxygen is added. In the latter case the addition of the oxygen results in the formation of

FIG. 90.   brown fumes, which are rapidly absorbed by the water, the volume of the gas diminishing as oxygen is added. In the former case no colouration results, and the volume of the gas is merely increased by the amount of nitrous oxide added. Care must be taken that the nitrous oxide is free from air.

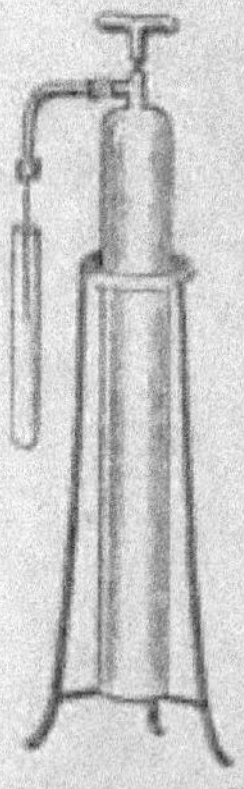

For liquefaction of nitrous oxide, see Liquefaction of Gases.

309. To collect liquid nitrous oxide from a cylinder of the condensed gas.[1] For this purpose a specially-constructed jet is employed, consisting of a fine steel capillary tube, attached to a wider elbow tube, which can be screwed to the metal cylinder or bottle. The fine jet is thrust 6 or 7 centimetres into the

[1] See Experiment 556.

open end of a glass tube closed at the bottom, and rather longer than a test tube, which should be held in a clamp. On gently releasing the cock of the bottle, a stream of the liquefied gas will be driven into the tube.

In sealing the end of the tube a very slight enlargement should be made by gently blowing out the closed end. The object of this is to make the lump of solid mercury (obtained in Exp. 310) rather larger than the rest of the tube, so that the tube may be inverted without the metal dropping out until it has partially melted.

310. To freeze mercury with liquid nitrous oxide. A few cubic centimetres of mercury may be poured into the tube containing the liquid collected as above; the mercury will at once solidify.

311. To show the combustion of charcoal upon the surface of liquid nitrous oxide. A fragment of charcoal about the size of a pea is ignited at one point in a gas flame, the charcoal being held by a pair of forceps, and then dropped into the tube containing the liquid. The charcoal bursts into active combustion, and dances about upon the surface of the liquid. This experiment may be done after the freezing of mercury, and while the solid metal is still in the bottom of the tube.

If the tube be now inverted over a plate, the mercury will be seen to be frozen into the tube, and if left for a few moments the solid lump will drop out upon the plate.

312. The composition of nitrous oxide may be shown by burning a fragment of phosphorus in a measured volume of the gas. This is conveniently done in the apparatus described on page 197 for showing the composition of carbon dioxide. The globe is filled with nitrous oxide by displacement, and a small fragment of phosphorus is placed upon the little spoon and lowered into the apparatus. The phosphorus is ignited by means of the wire heated by the passage of an electric current, and after it has combined with the oxygen, and the temperature has cooled, the volume of the residual nitrogen will be

L.

found to be the same as that of the nitrous oxide originally present.

313. The volume of both the nitrogen and the oxygen present in nitrous oxide may be shown by exploding equal volumes of nitrous oxide and hydrogen. Equal volumes of these two gases are introduced into the closed limb of a U-shaped eudiometer, in the manner described on page 41, and the mixture exploded as there explained, but without heating the tube with amyl alcohol vapour. On readjusting the level of the mercury the residual nitrogen will occupy half the volume of the mixed gases—that is, the same volume as the nitrous oxide originally present. The hydrogen, equal in volume to the nitrous oxide, will have united with half its own volume of oxygen to form water. Therefore the volume of oxygen present in the nitrous oxide is equal to half the volume of the nitrous oxide.

NITRIC OXIDE. NO.

314. **Preparation by the action of copper upon nitric acid.** $3Cu + 8HNO_3 = 3Cu(NO_3)_2 + 4H_2O + 2NO$. A quantity of copper clippings is placed in a two-necked Woulf's bottle, fitted with a thistle funnel and a delivery tube. Water to the depth of about one-fourth of the bottle is introduced, and about the same volume of strong nitric acid added. A rapid evolution of gas soon results, and all the gas required should be collected at once, for, as the reaction proceeds and the amount of copper nitrate in solution increases, gradually increasing quantities of nitrous oxide are produced. On this account it is well to collect all the gas in one vessel, and to test the purity of it roughly by taking a small sample of it in a cylinder and gradually adding oxygen to it. The oxygen combines with the nitric oxide, and the products of the combination dissolve in the water; the residual gas, therefore, represents the amount of impurity present in the gas.

315. Preparation by the action of sulphuric acid upon a mixture of potassium nitrate and ferrous sulphate. $2KNO_3 + 6FeSO_4 + 5H_2SO_4 = 2HKSO_4 + 3Fe_2(SO_4)_3 + 4H_2O + 2NO$. A mixture of the two salts, in the proportion of about four of ferrous sulphate to one of nitre, is placed in a small flask provided with a dropping funnel and a delivery tube. Water is added to the salts, and strong sulphuric acid dropped upon the mixture from the funnel, a gentle heat being applied to the flask. A steady evolution of very pure nitric oxide takes place.

316. To show the action of oxygen upon nitric oxide (see Nitrous Oxide, Exp. 308).

317. To show the behaviour of nitric oxide towards combustibles. A lighted taper introduced into the gas is extinguished.

318. A fragment of phosphorus, when feebly burning, is extinguished when plunged into nitric oxide. The phosphorus contained upon a deflagrating spoon is touched at one point by a warm wire, or blade of a knife, and instantly lowered into the gas, when it will be extinguished. On withdrawing the spoon and allowing the phosphorus to get into active combustion, and again lowering it into the gas, it will be seen to burn brilliantly.

319. To show the combustion of carbon disulphide vapour and nitric oxide. A small glass bulb containing a few cubic centimetres of carbon disulphide is dropped into a cylinder of nitric oxide and the plate quickly replaced. If the bulb is not broken by its fall, it may be readily fractured by being shaken forcibly against the bottom of the cylinder. The vapour at once diffuses itself into the nitric oxide, and on applying a lighted taper to the contents of the vessel combustion of the mixture results, accompanied by a bluish light of intense brilliancy, the light being particularly rich in actinic rays. It is convenient to

keep a supply of these bulbs containing carbon disulphide; the bulb is blown upon the end of a short piece of tube, having a long capillary stem. As each one is blown, and while still hot, the stem is dipped into a bottle of carbon disulphide, which as the bulb cools will be drawn up into it, and the stem can then be sealed off by a fine blowpipe flame.

The quantity of carbon disulphide to be used will obviously depend upon the capacity of the cylinder employed for the experiment, and should be arranged by a previous trial; it will be found that about 6 c.c. of the disulphide to a litre of the gas gives the best result.

320. The absorption of nitric oxide by a solution of ferrous sulphate, with the formation of the dark-brown colouration, may be shown by allowing a few bubbles of the gas to pass into a quantity of a solution of ferrous sulphate contained in a tall cylinder.

321. To show the composition of nitric oxide. This may be done by heating a spiral of iron wire to bright redness in a measured volume of the gas.

For this purpose a glass tube, about 50 centimetres long and 22 mm. bore, is fitted at one end with a cork through which pass two stout copper wires; by means of two small binding screws, the ends of these wires are connected together with a spiral of iron wire (fig. 91), the spiral being about 40 mm. long.

Two india-rubber rings are placed upon the tube, marking two equal volumes measured from the top, the space between the rings being about 18 centimetres. The spiral should reach to within about 2 centimetres of the upper ring, which will insure its being far enough from the cork not to over-heat it. The tube is filled with nitric oxide over water down to the second ring, the gas being obtained by the ferrous sulphate and nitre method (Exp. 315). An electric current is then passed through the spiral,

and the temperature of the iron gradually raised by regulating
the strength of the current; the wire should not be raised to
the point of visible redness until the walls
of the glass tube round it and above it are
quite warm, and free from wet upon the
inner surface. The current may then be increased until the wire shows a bright-red heat,
at which temperature it must be maintained
for 15 or 20 minutes. When the reaction
is complete and the whole allowed to cool,
the water should rise to the upper ring,
showing that the nitric oxide contains half
its own volume of nitrogen. To show that
the residual gas is nitrogen, the tube may
be removed from the pneumatic trough and

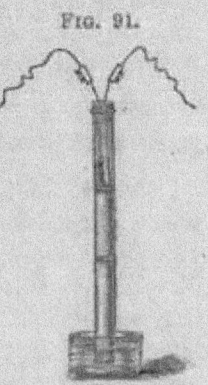

FIG. 91.

inverted, and the gas tested by means of a lighted taper; if the
decomposition has been complete, there will be no appearance
of red fumes on opening the tube to the air.

As iron wire is liable to contain appreciable quantities of
occluded hydrogen, it is well to heat the spiral for a short time
before using it, and to shake off the film of oxide which is
formed.

NITROGEN TRIOXIDE. N_2O_3?[1]

322. A few lumps of arsenious oxide are placed in a flask
fitted with a delivery tube, and a quantity of strong nitric acid
added. On gently warming the acid a rapid evolution of a
brown gas takes place. This gas is a mixture of nitric oxide
and nitrogen peroxide in varying proportions, resulting from
the following reactions, which go on simultaneously :—

$$As_4O_6 + 4HNO_3 = 2As_2O_5 + 2H_2O + 2NO + 2NO_2.$$
$$As_4O_6 + 8HNO_3 = 2As_2O_5 + 4H_2O + 8NO_2.$$

[1] There is considerable difference of opinion as to whether the compound
obtained by this reaction is nitrogen trioxide, the existence of N_2O_3 being a
matter of some uncertainty.

If the gaseous product from the action be passed through a gas-condensing tube (see Liquefaction of Gases, Exp. 649) immersed in a freezing mixture of ice and salt, or ice and calcium chloride, a blue liquid is obtained, which is believed to be the compound nitrogen trioxide, N_2O_3.

323. To show the formation of ammonium nitrite by the oxidation of ammonia (see Ammonia, Exp. 291).

324. To show that nitrous acid is both an oxidising and a reducing agent. A dilute solution of a nitrite is acidulated with dilute sulphuric acid and divided into two portions. A dilute solution of magenta is added to one portion; the magenta becomes oxidised and its colour discharged. To the other portion some dilute solution of potassium permanganate is added, which is reduced by the nitrous acid and its colour thereby destroyed. If the coloured solution in each case be contained in an ordinary wash-bottle, and poured in a thin stream into the acidulated liquid, the extreme rapidity of the reaction will be strikingly shown.

NITROGEN PEROXIDE. NO_2.

325. Preparation from lead nitrate. $Pb(NO_3)_2 = PbO + O + 2NO_2$. A quantity of lead nitrate, which has been roughly powdered and carefully dried in an oven, is introduced into a Wurtz flask, the branch tube of which is bent at right angles to the neck, and connected to a condensing tube immersed in a freezing mixture. The neck of the flask is closed with a cork, and a gentle heat applied to the salt; nitrogen peroxide is evolved, which condenses in the cooled tube. The liquid obtained will probably be blue (depending upon the degree of dryness of the lead nitrate). In this case a stream of dry oxygen should be passed through it while still cooled, the oxygen being admitted through a glass tube drawn out to a capillary tube. In a short time the blue colour will disappear, and the liquid will appear a pale straw colour.

326. Nitrogen peroxide may also be generated by acting upon tin with nitric acid. $5Sn + 20HNO_3 = Sn_5O_6(HO)_{10} + 5H_2O + 20NO_2$. A quantity of granulated tin is placed in a flask provided with a thistle funnel and delivery tube. Strong nitric acid is added, when a rapid evolution of gas results. The gas may be led through a small empty Woulf's bottle, and thence into a large dry flask, in which it can be collected by displacement.

327. To show the effect of heat upon nitrogen peroxide. A quantity of nitrogen peroxide is delivered into a flask, the neck of which has previously been drawn out before the blowpipe, and the neck is then sealed up. If the flask be cooled, the colour of the gas becomes less deep, approaching to a straw colour, and on warming the flask by means of a Bunsen the gas rapidly deepens in colour until a rich dark-brownish red is reached. This experiment may be made in front of the lamp, a ray of light being passed through the flask. As the flask is heated, the gradual darkening of the colour of the gas is seen by the disc of light upon the screen becoming more and more red, until finally it is nearly obscured.

NITROGEN PENTOXIDE. N_2O_5.

328. Preparation by the action of phosphorus pentoxide upon nitric acid. $2HNO_3 + P_2O_5 = 2HPO_3 + N_2O_5$. A quantity of phosphorus pentoxide is introduced into a small tubulated retort, and fuming nitric acid (sp.gr. 1·5) added until the mixture is of the consistency of thick paste. The contents of the retort are then gently heated, when nitrogen pentoxide distils off, and may be condensed as a white crystalline substance in a receiver placed in a freezing mixture.

NITRIC ACID. HNO_3.

329. Preparation from nitre and sulphuric acid. $KNO_3 + H_2SO_4 = HKSO_4 + HNO_3$. Equal weights of nitre and sulphuric acid are introduced into a non-tubulated retort. About

60 grams of nitre in crystals are first placed in the retort, and the same weight of sulphuric acid poured upon it by means of a funnel with a long stem, reaching right down the neck of the retort in order that the acid should not touch the sides. A gentle heat is applied, and the nitric acid which distils over is condensed in a flask, which is cooled by being immersed in a dish of cold water. In this reaction the acid which is distilled is nearly colourless, and only a slight appearance of brown fumes is seen within the retort.

330. To show the preparation of nitric acid, using an excess of nitre, and thereby forming the di-potassium sulphate, and causing the decomposition of a portion of the acid into oxygen and nitrogen peroxide.

$$2KNO_3 + H_2SO_4 = K_2SO_4 + 2HNO_3$$
$$2HNO_3 = H_2O + 2NO_2 + O$$

Sixty grams of nitre are introduced into a retort, and half that weight of sulphuric acid added as described above. The neck of the retort is fitted by means of a ring of caout-chouc into a small tubulated receiver, the tubulure of which carries a delivery tube dipping into a pneumatic trough. On the first application of heat, nitric acid distils over, much as in the above experiment,

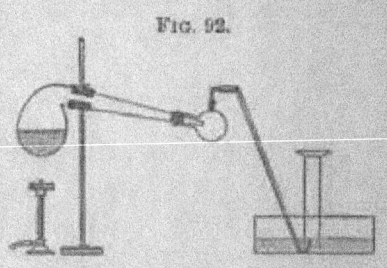

Fig. 92.

but as the operation proceeds the mass in the retort thickens, brown fumes are rapidly disengaged, along with oxygen, which can be collected over water, in which the nitrogen peroxide will all dissolve.

331. The decomposition of nitric acid by heat into nitrogen peroxide and oxygen may be strikingly shown in the following manner. A long clay tobacco-pipe is arranged as shown in the figure, with the mouthpiece just dipping beneath the water of the trough. The stem is made red hot at one point by means

of a Bunsen, and a little strong nitric acid is poured into the
bowl. The first drops that reach the heated spot are decomposed,

FIG. 92A.

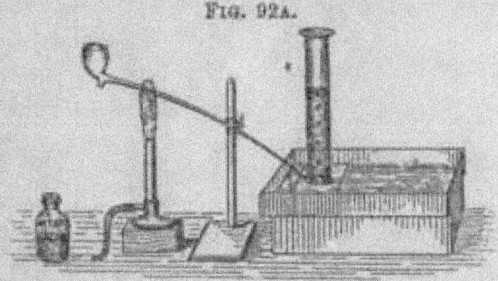

and the gaseous products, finding less resistance to their exit
through the small depth of water in the trough than through
the acid above them in the bowl, bubble steadily out by the
long stem.   The peroxide is dissolved by the water in the
trough, but the oxygen can be collected in a cylinder.

332. The oxidising power of nitric acid may be illustrated
by pouring a small quantity of the fuming acid (sp.gr. 1·5)
upon a quantity of sawdust contained in a platinum dish,
which has been gently heated over a rose burner until the
wood has become dry and just begun to char. On pouring a
small quantity of the acid upon it, it at once bursts into flame.
Further illustrations may be seen in the action of nitric acid
upon arsenious oxide, phosphorus, and various metals.

333. To show the oxidising power of potassium nitrate. A
quantity of nitre is heated in a Florence flask to a temperature
well above its melting-point, and a small fragment of charcoal
about the size of a pea, which has been just ignited at one point,
is dropped in; the charcoal at once bursts into flame, and
dances about upon the surface of the melted salt.

In like manner a fragment of sulphur, without being previously ignited, may be dropped into melted nitre, when a very
brilliant combustion takes place. In this case the nitre may
be melted in a test tube, there being no gain in the large surface of melted salt, as in the case of the carbon.

THE ATMOSPHERE

334. To show the suspended impurities in the air, a parallel beam of electric light (see Lantern Illustrations) is allowed to pass through the air of a darkened room. The track of the beam will be rendered visible by the suspended matter floating about, much of which, when thus strongly illuminated, may be distinctly seen as an infinite number of distinct particles.

335. To show that a ray of light, passing through air from which the suspended impurity has been removed, is not visible. An oblong-shaped glass box is made by binding pieces of glass together by means of strips of brown paper pasted along the edges. Before finally closing it up, a small quantity of glycerine is smeared over the bottom, after which the box is completed, and allowed to stand for some hours or a day or two. The suspended matter gradually settles and sticks to the glycerine.

If this box be held in the beam, the track of the ray as it traverses the air within the box is perfectly invisible; its path

FIG. 93.

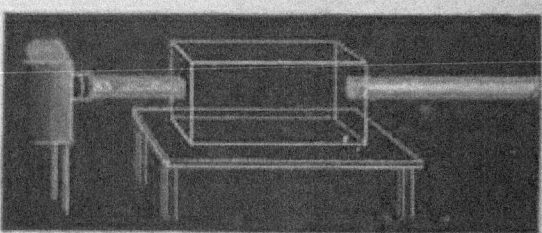

being only visible in the air of the room as it approaches and leaves the box (fig. 93).

336. To show that solid matter can remain suspended for a long time in the atmosphere. For this purpose a number of Bunsen lamps are lighted and placed in various parts of the room, and a quantity of sodium oxide discharged into the atmosphere. This may be conveniently done by placing a fragment of sodium upon a wet piece of filter paper in a plate; the metal almost immediately takes fire, producing a cloud of smoke of the oxide. A bell-jar should be in readiness to cover the burning sodium before its combustion is completed, in order to prevent the fused globule of caustic soda which is formed from being thrown about. As the cloud of oxide diffuses itself throughout the room, the Bunsen flames become infected, and show the characteristic yellow produced by sodium compounds.

337. To show the effect of heat upon the suspended matter in the air. A Bunsen flame is held immediately below the beam

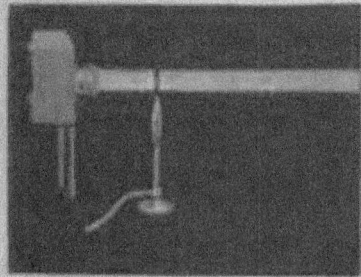

FIG. 94.

of electric light. The beam for this purpose may be made to converge, and the lamp placed below it at a point where it is concentrated. There will be seen rising through the beam what appears like black smoke, but what is in reality air which has been so far deprived of its suspended matter, by the heat of the flame, that it is unable to reflect light, and therefore the beam is at that point invisible, just as in the case described in Experiment 335.

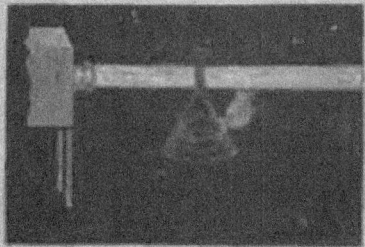

FIG. 95.

Instead of a flame, a similar effect may be produced by holding a hot poker, or, better, a soldering bit, beneath the ray, but in this case the effect is less pronounced.

338. A large proportion of the suspended matter, presumably the liquid portions, may be removed by the heat of a flask of hot oil. If a quantity of olive oil be heated in a flask, and the flask be brought immediately beneath the beam of light,

especially if a cone of paper be placed over the flask in order to concentrate the rising current of heated air, the same appearance may be seen.

339. The suspended matter may be filtered out of the air by means of cotton wool. A wide glass tube, about 30 centimetres long, is packed with cotton wool. One end of the tube is fitted with a cork with a smaller tube, to which a piece of caoutchouc can be attached. To prevent the wool from being blown out of the tube when air is sent through it, a ring of cork should be pushed into the open end. If a gentle current of air be sent through the tube, and the filtered air be passed into the beam, it will be seen to show the same blackness as in the cases above described.

For this experiment the air may be blown through the tube from the lungs, or a foot bellows may be used; in the latter case the stream of air should be regulated by a screw clamp upon the caoutchouc tube, and not by the bellows themselves.

340. It may be shown that the last portions of air exhaled from the lungs in the act of respiration are in a like manner deprived of a portion of the suspended matter. A long expiration is made through a glass tube into the narrow part of the beam; at first the exhaled air appears as full of suspended matter as the surrounding atmosphere, but as those portions of air are discharged which have been longest in contact with the lungs, they will be seen to produce a similar effect upon the beam as the air which has been filtered through cotton wool. It is well to slip a short piece of caoutchouc tube over the end of the tube which is brought near the beam of light, in order to prevent any light from being reflected from the edges of the tube, which in this experiment will be brought up close to the beam, as any scattering of light from this cause would make it difficult to see the dark stream of air ascending through the beam.

341. To show the influence of suspended matter upon the formation of fog. A large 'bolt-head' flask is fitted with a caoutchouc cork, carrying two short tubes, which pass just

through the cork. One of these is connected to an air-pump by means of a piece of wire-lined caoutchouc tube; to the other tube is attached a wider glass tube, about 45 centimetres long, which is packed with cotton wool, a screw clamp being placed upon the short connecting piece of caoutchouc tube which unites this filtering tube to the flask. The inside of the flask is moistened by introducing a small quantity of water, and the cork inserted. The apparatus may then be allowed to stand for at least twenty-four hours, communication with the outer air being closed, when the suspended matter in the air within the flask will have settled; or the flask may be exhausted by means

FIG. 96.

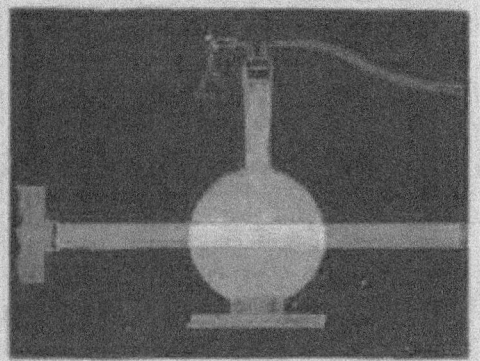

of the pump, and air allowed very slowly to re-enter through the filter tube. By repeating this operation once or twice the air in the flask will be almost entirely deprived of suspended matter, and when placed in the path of a strong beam of light, the absence of the suspended matter may be seen by the invisibility of the beam as it traverses the flask. If, while the air is in this condition of comparative freedom from suspended impurity, one or two rapid strokes of the pump be made, the cooling effect of the sudden expansion of the air within the flask will result in the deposition of moisture (the air being saturated), and it will be seen that the moisture is deposited in the form of a fine rain, distinctly visible in the strong light; it will also be noticed that this fine shower of rain is hardly more than of momentary duration. If the screw clamp be now opened, and air allowed slowly to enter through the filter, to replace that which was removed by the pump, the same experiment may be once more performed. If now the filter tube be disconnected, and unfiltered air from the room be allowed to enter the flask, even in very small quantity, and the experiment once more repeated, a dense fog will immediately make its appearance in the flask, which will persist for a considerable time.

If a minute trace of coal smoke, or smoke from burning sulphur, be allowed to enter the flask (which may be done by replacing the filter tube by an ordinary funnel, and holding this for a moment over the burning material so as to catch a little of the products of combustion in the inverted funnel, and then opening the screw clamp in order that they may be drawn into the flask), and the air once more rarefied, it will be seen that the fog that is now produced will be much more permanent than the former.

342. The fact that air is ponderable may be demonstrated by weighing a flask filled with air, and the same flask from which the air has been exhausted.

A round and moderately thick glass flask has fitted into its mouth, by means of a caoutchouc cork, a brass stopcock which can be screwed to an air-pump. The flask is then partially exhausted and counterpoised upon a balance; on opening the cock and allowing air to enter, the balance will show the increase of weight due to the entering air. With a moderately delicate balance (such as that described under Hydrogen, Exp. 13), the weight of the air which can be easily withdrawn from a flask by the lungs alone can readily be shown. An ordinary flask of about two litres capacity is fitted with a cork carrying a glass stopcock; a portion of the air within the flask is sucked out, and the flask counterpoised. On opening the cock the increase of weight by the re-entering air will be seen by the descent of the arm of the balance which carries the flask.

343. The pressure exerted by the atmosphere may be shown by withdrawing the air from beneath a stretched membrane, either of india-rubber, or of bladder. A thin sheet of rubber or a piece of bladder is tied over the mouth of a short glass cylinder, which is ground to fit upon an air-pump plate. As the air is exhausted from below the membrane, and the weight of the atmosphere above begins to be felt, in the case of the rubber, it will be pressed down into the cylinder more and more, until it finally breaks; while with the bladder the rupture is sudden and almost explosive.

The 'Magdeburg hemispheres,' the 'sucker,' and a number of other instruments, may be used to illustrate the same fact.

344. The effect of the pressure of the atmosphere in supporting a column of mercury may be shown by completely filling a

Fig. 97.

glass tube about 900 millimetres long with mercury, and inverting the tube into a trough containing mercury. By surrounding the trough with an air-pump receiver and gradually exhausting the air, the column of mercury in the tube will be seen to fall until, when the air is practically all withdrawn, the mercury in the tube will be level with that in the trough; on allowing air to re-enter, the mercury once more ascends to its former height of about 760 millimetres. The mercury should be contained in a beaker, large enough to hold all that is in the long tube without overflowing; a small tubulated bell-jar is lowered over the tube, and fitted to it by means of a caoutchouc cork previously slipped over the tube.

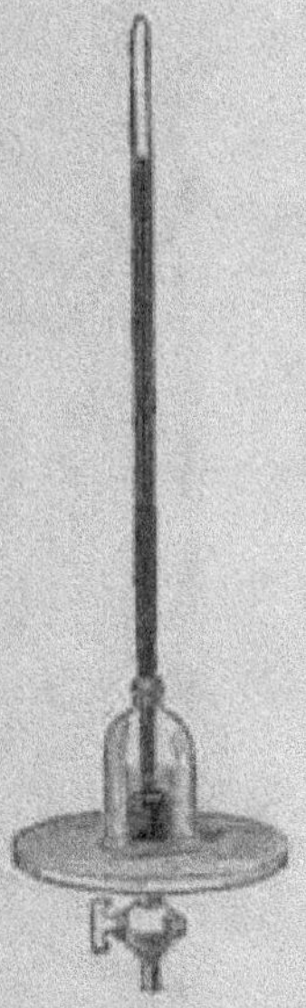

345. To show the volume composition of the air by means of phosphorus. (See Nitrogen, Exp. 255.)

346. To show the volume composition of air by means of an alkaline solution of pyrogallol. A graduated eudiometer, bent as shown in fig. 98, with the short limb of rather wider tube than the rest, is filled with air nearly down to the bend; the eudiometer is then lowered down into a tall cylinder of water in which a thermometer is suspended. When the tube has acquired the temperature of the water it is raised by means of a string attached to an india-rubber ring upon the tube, until the level of the water in the tube is the same as that in the cylinder, and the volume of the gas read off. It is then withdrawn from the cylinder, and by means of a pipette as much of the water is removed from the open limb as can be taken without allowing air to enter the eudiometer; the limb is then half filled with a strong solution of sodium or potassium hydroxide, and then filled up nearly to the top with a solution of

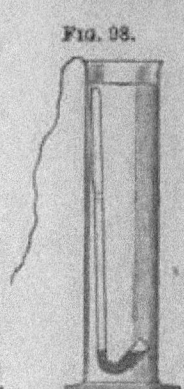

Fig. 98.

pyrogallol, and a caoutchouc stopper inserted in such a way that no air remains between it and the liquid. The inclosed air is then agitated with the solution for a few moments, when the stopper is removed under water in a trough or basin and the dark liquid allowed to flow out. The tube is then replaced in the cylinder, and the volume of the gas read off with the same precautions as before.

347. To show the presence of carbon dioxide in the air see Carbon Dioxide, Exp. 400.

348. To show the presence of aqueous vapour in the air. A glass vessel in which a freezing mixture is contained is seen to become rapidly covered with hoar frost. If it be desired to collect liquid water condensed from the air, a round-bottomed flask may be filled with broken ice, and suspended over a glass dish or beaker; the condensed aqueous vapour will run down the sides of the flask and drop into the beaker below.

349. The aqueous vapour present in the air may also be shown by aspirating air through U tubes filled with desiccating

agents, and which have been counterpoised upon a balance. Three U tubes filled with pumice, moistened with sulphuric acid, are joined in series and counterpoised. They are then connected to an aspirator, and a stream of air drawn through them for a short time. On again disconnecting them from the aspirator they will be found to have increased in weight. This experiment may be made roughly quantitative by aspirating a known volume of air through the tubes, either by means of a 'swivel aspirator' (fig. 99), or by emptying a vessel of known capacity filled with water, the absorbing tubes being weighed before and after the measured volume of air has been passed through; it is advisable in this case to introduce into the series a small tube containing phosphoric pentoxide.

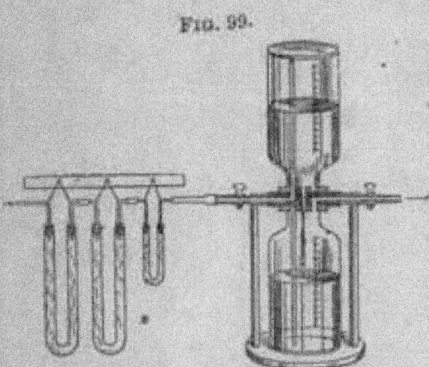

FIG. 99.

DIFFUSION

350. To show that a light gas, as hydrogen, will escape from a vessel which is held with its mouth downward. A small cylinder is filled with hydrogen, and supported in an inverted position. The plate is removed, and the vessel allowed to remain undisturbed for a certain time, at the expiration of which a lighted taper is applied to the cylinder; the gas will have become so far replaced by air that the combustion will be attended with the characteristic sound produced by combustion of mixtures of air and hydrogen. It may be compared with the effect produced by igniting hydrogen in a similar cylinder which

has been filled with the gas. The length of time necessary to insure a detonating mixture of air and hydrogen must be determined by previous trial. A cylinder 30 millimetres diameter and 12 centimetres long must be allowed from 2 to 3 minutes.

351. To show the escape of hydrogen through a cracked flask. A flask having a minute crack in it is fitted with a cork carrying a straight piece of small bore tube about 20 centimetres long. The flask is filled with hydrogen by displacement, the cork inserted, and the apparatus inverted and supported in a clamp so as to dip the open end of the tube into a small beaker of coloured water. As the hydrogen diffuses through the crack the coloured liquid will be seen to rise in the narrow tube. The flask for this experiment may be cracked by touching it at one spot near the bottom with the point of a hot piece of glass, and then quickly touching the same spot with a moistened finger; this usually results in the production of a short crooked crack. It may be necessary to crack two or three flasks before one is obtained that is satisfactory, but when one is so secured it should be carefully preserved; it is well to keep it under a glass bell-jar in order to prevent its getting dirty, as it should on no account be washed.

352. To show the diffusion of gases into each other through a long tube. Two soda-water bottles are connected together by a piece of combustion tube about 60 centimetres long, the tube being fitted to the bottles by corks. One of the bottles is filled with hydrogen and the other with oxygen, and the apparatus secured in a vertical position with the hydrogen uppermost. At the expiration of a few hours it will be found that sufficient of the hydrogen will have diffused down into the oxygen bottle to cause the mixture to detonate on the application of a flame to the bottle; and in the same way oxygen may be shown to have risen into the hydrogen bottle. It is advisable to prepare this experiment the day before it is required, as 24 hours is not too long a time to allow for the gases to become thoroughly mixed.

353. To show the diffusion of hydrogen through porous material (Graham's diffusiometer). A bulb of about 200 cubic centimetres capacity is blown upon a glass tube about 70 or 80 centimetres long and 7 millimetres bore. The bulb has also a short neck, about 25 millimetres wide, into which is fastened a thin disc of porous material. The most convenient substance for the purpose is a flat piece of a porous cell used in a Grove's battery. A circle the size required is marked upon a fragment of such a pot, and the piece roughly nibbled round with a pair of pliers, after which it is filed to the exact size. It is fixed into the neck, about half-way down, by means of cement, and the neck is then closed with a caoutchouc stopper. The apparatus is filled with hydrogen by displacement, and supported with the end of the long tube dipping into coloured water in a glass basin. On withdrawing the caoutchouc stopper, hydrogen quickly diffuses out through the biscuit plug, and the coloured water will rapidly rise in the tube.

FIG. 100.

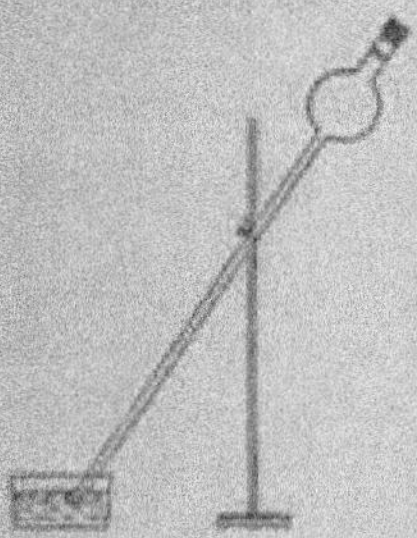

354. A convenient apparatus for demonstrating the rapidity of diffusion may be constructed as follows: A round porous cell, such as is used in a Bunsen's battery, is fitted by means of a cork to one limb of a tall U tube, containing coloured water, the whole being conveniently mounted on a wooden stand as seen in fig. 101. On bringing a beaker of hydrogen over the porous cell, the level of the liquid in the two limbs is instantly disturbed, the water being almost immediately driven to the top of the open limb of the tube. On removing the beaker, the hydrogen which had entered the apparatus now diffuses out, and so much more rapidly than air makes its way in, that the liquid will be drawn up towards the cell, some distance above the level.

By cementing a perforated glass plate to the under side of the cork, and standing upon it a bottomless beaker, the porous cell may be surrounded with carbon dioxide, by pouring a vessel filled with the gas into the beaker. A disturbance of the level of the liquid in the U tube will be seen to follow, but in the

FIG. 101.   FIG. 102.

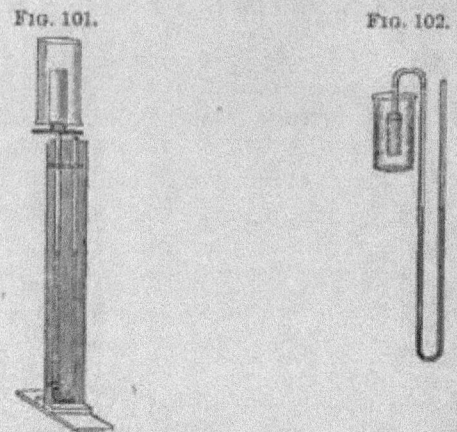

opposite direction and to a less degree than that which resulted when hydrogen gas was used.

The apparatus for these experiments may, if desired, be modified, and constructed as shown in fig. 102, where the porous vessel is connected to the U tube by means of a short piece of flexible tube, so that it may be either introduced into an inverted jar of hydrogen or dipped into a vessel of carbon dioxide.[1]

FIG. 103.

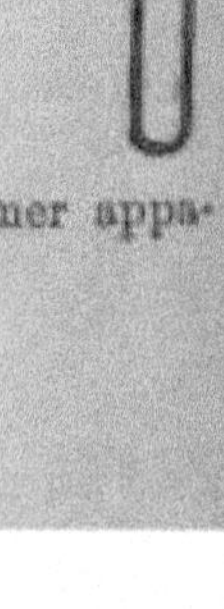

355. On a small scale these experiments may be performed with a piece of apparatus in which a clay tobacco pipe is the porous portion (fig. 103). The mouth of the bowl is closed by cementing over it a little disc of porous material (see Experiment 353), and the stem broken off about 8 centimetres from the bowl. By means of a short piece of caoutchouc this is attached to a small manometer, as in the former apparatus, containing coloured water.

[1] See Table XX. in the Appendix.

356. By a slight modification of the above experiment a fountain of coloured water may be projected into the air to a considerable height.

The porous vessel is attached to the end of a glass tube about a metre long, which passes through a cork into a small two-necked Woulf's bottle, the tube reaching nearly to the bottom of the bottle. Into the other neck is fitted a glass tube drawn out to a jet, and which also reaches nearly to the bottom, the bottle being nearly filled with coloured water. On bringing a beaker of hydrogen over the porous pot the gas which passes in causes air to bubble into the Woulf's bottle, and the pressure which is produced drives the water up the jet, from which it is projected to a height of about half a metre.

FIG. 104.

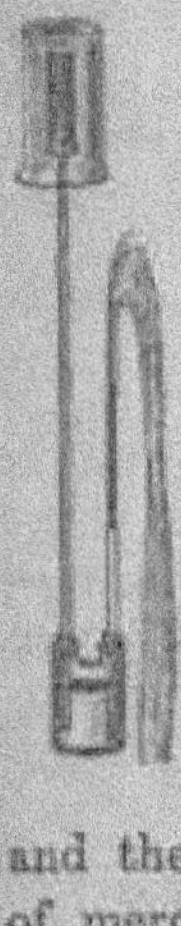

357. To illustrate the principle of Ansell's fire-damp indicator. A piece of apparatus similar to that shown in fig. 101, but conveniently rather smaller, has two platinum wires fused into the U tube, one in the bend, and the other some distance up the open limb. A quantity of mercury is placed in the tube, so that when it is at the same level in the two limbs the upper platinum is three or four millimetres above the surface. The instrument is placed in an electric circuit with a bell. When a vessel of marsh-gas (or coal-gas) is brought over the porous pot the mercury will be driven up the open limb, and, coming in contact with the platinum wire, will close the electric circuit and cause the bell to ring. The bell may be placed at any convenient distance from the apparatus.

358. To show gaseous diffusion through a soap film. A few cubic centimetres of ether are placed in the bottom of an inverted bell-jar or large beaker, which may be covered with a piece of cardboard. A soap bubble[1] blown upon a thistle funnel is lowered for a second or two into the ether-laden atmosphere

[1] For soap solution see Experiment 14.

in the bell-jar and again withdrawn; the tube being closed with the thumb. A lighted taper is then applied to the open end of the glass tube, and as the bubble slowly collapses and expels its contents a flame of burning ether vapour will be obtained.

FIG. 105.

FIG. 106.

359. To show the separation of gases in a mixture by diffusion (Atmolysis). Two long clay tobacco pipes are connected together by a short piece of caoutchouc tube upon their mouth-pieces. The bowls are fitted with corks, so that a stream of gas can be passed through the pipes and led into a pneumatic trough. The pipes may be conveniently arranged as shown in the figure. A mixture of hydrogen and oxygen (2 to 1), contained in a small gas-holder, is caused to pass through the pipes, and, by regulating the rate at which they pass, oxygen gas, practically free from hydrogen, may be collected at the trough, and

FIG. 107.

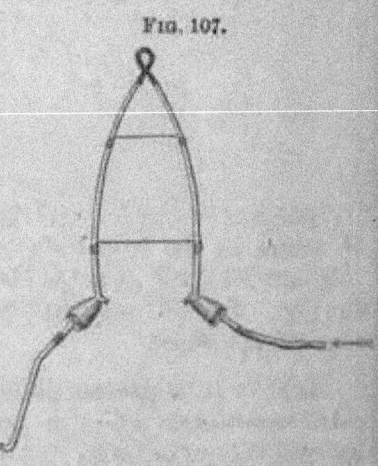

the gas shown to be oxygen by its action upon a glowing splint. The exact rate at which the gases should be passed through in order to give this result must be ascertained by a trial, as it will differ with different pieces of apparatus: with two ordinary 'churchwardens' the rate should be such that 100 cubic centimetres of gas are collected in about two minutes. If the gas be passed through too fast, that which is collected will still detonate; if too slowly, so much air will have become mixed with the oxygen that the gas will not rekindle a glowing splint. A cylinder of the gas contained in the gas-holder may be collected direct, without passing through the pipes, in order to show that the mixture employed is explosive.

360. Another example of the same order may be seen in the separation of ammonia from hydrochloric acid gas when ammonium chloride is undergoing dissociation. (See Dissociation, Exp. 637.)

COMBUSTION

361. To show that the terms 'combustible' and 'supporter of combustion' are relative, a number of experiments may be arranged.

Coal-gas and air. An ordinary paraffin or Argand lamp chimney is fitted at the bottom with a cork, through which pass two glass tubes. One is a straight tube about 10 centimetres long and 1 centimetre bore, which passes through the centre of the cork; the other is a piece of smaller tube, bent at right angles, which is connected with the coal-gas supply. A piece of asbestos card 6 or 7 centimetres square, having a round hole in the middle of about 1½ centimetre diameter, is placed upon the top of the chimney: a stream of coal-gas is passed through the apparatus, the hole in the card being loosely closed by laying upon it a piece of card or mica. In a few moments the vessel will be full of coal-gas, and the excess will escape by the straight central tube. A lighted taper is then applied to the issuing gas, and at the same moment the hole at the top is uncovered. The flame will then recede up the tube, drawing air after it, which will continue to burn with a feebly luminous flame in the surrounding atmosphere of coal-gas. The excess of coal gas escaping from the top may be ignited, so that two flames are produced; the one a flame of air burning in coal-gas, the other a flame of coal-gas burning in air. If, when the outside flame of coal-gas is not burning, a paper spill be thrust down from the top of the chimney, it will burn so long as it is held within the flame of burning air; but itself cannot be inflamed, and its combustion ceases as soon as it is removed from the flame.

FIG. 103.

362. The combustion of air in coal-gas may also be shown by filling a glass cylinder with coal-gas and inflaming the gas at the mouth while the cylinder is held in an inverted position. A glass tube delivering air is then passed up into the cylinder, and as it passes the burning coal-gas at the mouth the air is inflamed, and continues its combustion within the cylinder until the coal-gas is all consumed.

363. With hydrogen and air. This may be done exactly as described in No. 361 for coal-gas. In this case it will be noticed that the two flames much more closely resemble each other than when coal-gas is employed.

364. With carbon monoxide and air. This is also shown by means of the same apparatus, and the two flames in this case appear absolutely similar to each other.

365. With hydrogen and oxygen. An inverted glass cylinder of hydrogen is inflamed, and a jet of oxygen passed through the flame into the interior of the vessel, when the oxygen will continue to burn in the atmosphere of hydrogen until the latter gas is exhausted. For comparison a jet of burning hydrogen may be lowered into a cylinder of oxygen.

For this experiment the jets may be ordinary fish-tail burners screwed into pieces of lead or ' compo ' pipe.

FIG. 109

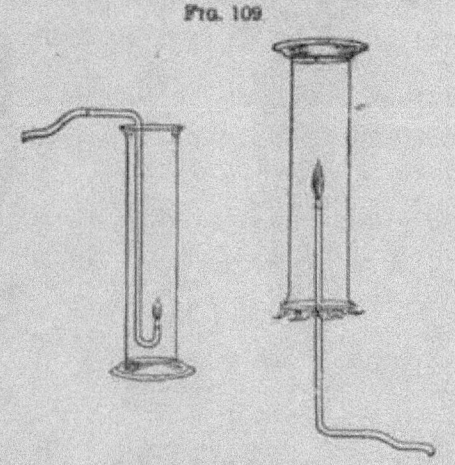

366. A striking variation of this experiment may be made by generating the oxygen for the combustion from strontium chlorate. A deflagrating spoon has the cup reversed, and a quantity of strontium chlorate is heated in it until it begins to give off oxygen; the spoon is then introduced up into an inverted cylinder of hydrogen, which has been inflamed at the mouth. As the heated chlorate passes the hydrogen flame, the oxygen it is evolving becomes ignited, and continues to burn with a brilliant red flame imparted by the strontium salt.

367. A continuous flame of oxygen burning in hydrogen may be obtained by a slight modification of the apparatus described for Experiment 361. In this case the short wide tube is replaced by a long piece of tube which fits sufficiently easily into the cork that it can be pushed either up or down. Into the upper end of this tube is inserted a little roll of platinum foil, which forms a short platinum tube projecting a few millimetres above the glass, and serves as a jet. A stream of hydrogen is delivered into the apparatus by the short tube H, and the gas

inflamed at the top as it issues through the hole in the asbestos card. The long tube is pushed up until its end is just within the hydrogen flame, and a small stream of oxygen passed up. The oxygen will ignite inside the hydrogen flame, and the tube may then be drawn down into the wide part of the apparatus where the flame of oxygen continues to burn. To extinguish the flame the tube is again pushed up to the top, and the oxygen supply then stopped. The hydrogen flame may be put out by cutting off the supply, and at the same time laying a piece of card upon the hole from which the flame was burning.

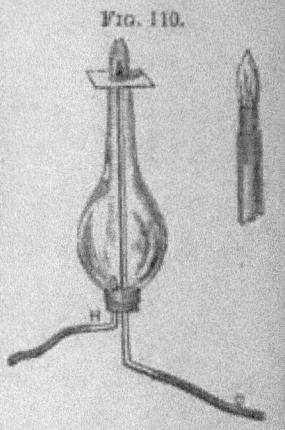

Fig. 110.

368. **With carbon monoxide and oxygen.** This is done in an apparatus precisely similar to that described above, in Experiment 367.

369. **With sulphuretted hydrogen and either air or oxygen.** The same as above.

370. **With ammonia and oxygen.** This also may be shown in the same apparatus, the stream of ammonia being generated by gently warming the strong aqueous solution. As ammonia will not burn in air, no flame will be obtained until the oxygen is passing, when the combustion of the ammonia will be seen to proceed round the oxygen jet. When the oxygen jet is drawn down into the interior of the chimney, it continues to burn in the ammonia gas, and sufficient hydrogen will be produced by the decomposition of a portion of the ammonia to render the gas escaping from the top inflammable, where it burns with the usual brownish tinge produced by ammonia.

In all the above experiments nitrous oxide may be substituted for oxygen.

371. With hydrogen and chlorine. The combustion of chlorine in an atmosphere of hydrogen may be demonstrated in the same manner as described for oxygen in hydrogen, Exp. 367.

(See also Hydrochloric Acid, Synthetical Formation, Exp. 182.)

372. With oxygen and sulphur vapour. A quantity of sulphur is boiled in a Florence flask, and the vapour inflamed as it issues into the air. A glass delivery tube having its end so closely recurved as to enable it to pass down the neck of the flask, is connected to an oxygen supply, and a small stream of gas passed through; as this is lowered through the flame of burning sulphur it ignites, and continues to burn brightly within the atmosphere of sulphur vapour.

373. To show the increase of weight resulting from combustion of substances in air.

By the combustion of reduced iron. A quantity of ferric oxide is reduced by being heated in a combustion tube in a stream of either hydrogen or coal-gas. When cold the black powder may be transferred to a stoppered bottle, in which it may be preserved.

A small heap of the reduced iron is placed upon a clean sand-bath and counterpoised upon a balance. The heap is then ignited by means of a taper, and as it continues to smoulder its increase in weight will be indicated by the balance.

The experiment may be varied by suspending a magnet upon the balance, the finely divided iron being held by the magnet; in this case care must be taken that what falls from the magnet during the combustion is caught upon the pan of the balance.

374. By the combustion of magnesium. Three strands of magnesium ribbon are twisted together, and suspended in a large inverted beaker in the manner indicated in the figure. A cork carrying a copper wire bent into a small hook at one end, is cemented on to the side of the beaker at a short distance from the bottom, the wire projecting beyond the middle of the beaker so that the heat from the magnesium as it burns may not soften the cement. The lip of the beaker should be ground smooth upon emery cloth moistened with turpentine, in order that it may stand quite close upon the ground glass plate P. This plate is perforated, and carries a short piece of rather wide tube T, which may be loosely packed with cotton wool, and it has cemented upon it three short pieces of wood which serve as legs.

FIG. 111.

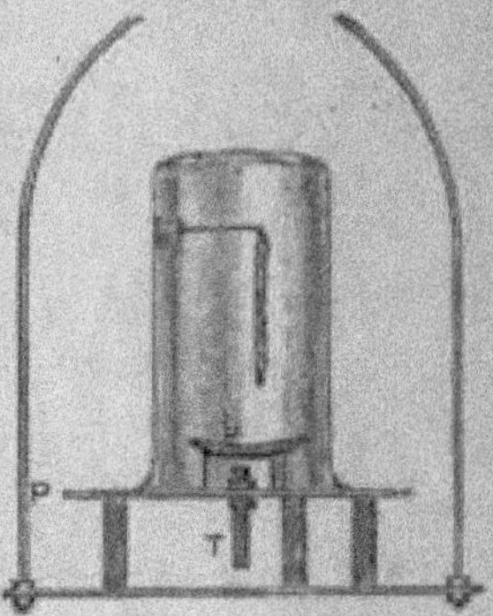

Supported by a small wire tripod is a clean sand-bath B, to receive the magnesium oxide which falls.

The whole apparatus is placed upon the pan of a balance and counterpoised; the beaker is then lifted, and the magnesium ignited by a small gas or spirit-lamp flame; the beaker is quickly replaced, and the combustion allowed to proceed until the whole of the metal is consumed. If the beaker fits close upon the plate, little or no oxide will escape beneath the lip, and the cotton wool in the tube T will serve as a filter for retaining the particles which might otherwise escape with the air which is expelled by the heat resulting from the combustion; on allowing the vessel to cool, and again liberating the balance, it will be seen that an appreciable increase in weight has taken place.

375. By the combustion of phosphorus. A flask of about 2 litres capacity is fitted with a cork carrying a glass stopcock and two stout wires, which reach rather below the centre of the flask. Upon the end of one wire is attached a deflagrating spoon, and the two wires are connected by means of a loop of thin platinum wire, which is made to dip down into the spoon without actually touching it, so that when it is heated by an electric current, it will cause the ignition of a piece of phosphorus.

A small fragment of phosphorus is placed upon the spoon, the stopcock closed, and the flask counterpoised. The phosphorus is then ignited by the passage of the current, and allowed to burn itself out, the stopcock being still closed. It may be shown that under these circumstances no change of weight has taken place. On allowing the flask to cool, and opening the cock, the sound of inrushing air will be heard, and on again weighing the vessel the increase of weight will be manifest.

376. By the combustion of sulphur. A glass tube 24 centimetres long and about 4 centimetres wide (a common Argand lamp glass may be used) has fitted into it a wooden cork perforated with two or three holes. A shallow platinum capsule is fastened to a copper wire, and fixed into the cork as shown in the figure. A false bottom made of a piece of iron-wire gauze is placed in the cylinder, about 15 centimetres from the top. It may be secured in position by means of fine copper wires passing over the top and out from below, and which are twisted together on the outside of the glass; three such wires will keep the piece of gauze in its place. These wires also afford a means of hanging the apparatus, by attaching to them three short pieces of wire which are twisted together to form a loop. A quantity of caustic soda, either in lumps or in sticks, is placed upon the gauze until the tube is loosely filled to within a short distance of the top. A fragment of sulphur is placed upon the capsule, to which it may be melted by gently warming, and the whole apparatus counterpoised. The cork is then withdrawn, and the sulphur ignited and the cork again quickly replaced. As the sulphur continues burning, the whole apparatus will be seen to increase in weight, notwithstanding that some of the products of combustion are escaping from the cylinder in the form of slight fumes.

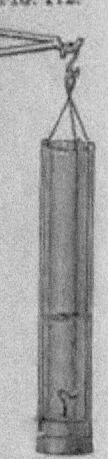

FIG. 112.

377. By the combustion of an ordinary candle. For this purpose the apparatus above described may be employed, the wire carrying the platinum capsule being removed, and a short piece of candle placed upon the cork in its stead. The candle may be melted to the cork.

378. The alternate reduction and reoxidation of copper oxide, and its alternate loss and increase in weight in the process, may be shown in the following way. A quantity of copper oxide is moistened with a strong solution of gum until it has the consistency of dough, and a quantity of it is plastered upon a thick copper wire until a rod of the oxide is obtained about the thickness of the finger, and about 12 centimetres long, the copper wire projecting about 15 centimetres beyond the oxide as a convenient means of holding it. Several of these may be made and hung up for a few days in a warm place to dry. When perfectly dry a thin copper wire is wound round the oxide, the turns being about 5 millimetres apart, and the mass carefully heated by means of a Bunsen until the organic matter is destroyed. It is then suspended from the beam of a balance, as shown in the figure, by means of thin wire, a small weight being hung upon the end to keep it in position. The whole is counterpoised, and the copper oxide heated, and a glass adapter through which hydrogen is streaming is brought over the hot mass. The oxide at once begins to glow in the hydrogen, and continues burning until the whole is reduced. The water formed for the most part escapes as steam, although a portion of it at first may condense upon the glass and drop from the end; it is well, therefore, that the pan of the balance be removed as seen in the figure, so that any drops

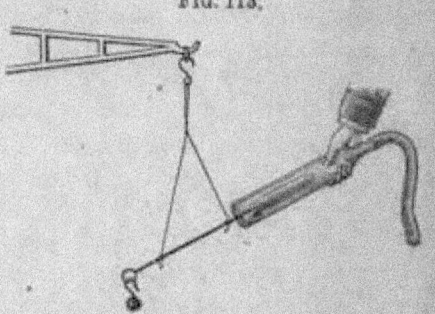

Fig. 113.

which may so fall shall not interfere with the experiment. During this reduction of the oxide the loss of weight will be seen by the equilibrium of the system being disturbed.

As soon as the combustion of the oxide begins to cease the tube is withdrawn and quickly replaced by a similar one through which oxygen is passing. The combustion at once recommences, the reduced copper burning with so much energy that the entire mass becomes incandescent, and at the same time the material regains its original weight.

379. Ignition point, or the temperature to which a substance has to be raised before combustion commences. The wide difference existing between the igniting points of various substances which burn in the air may be illustrated by the following experiments:

Fig. 114.

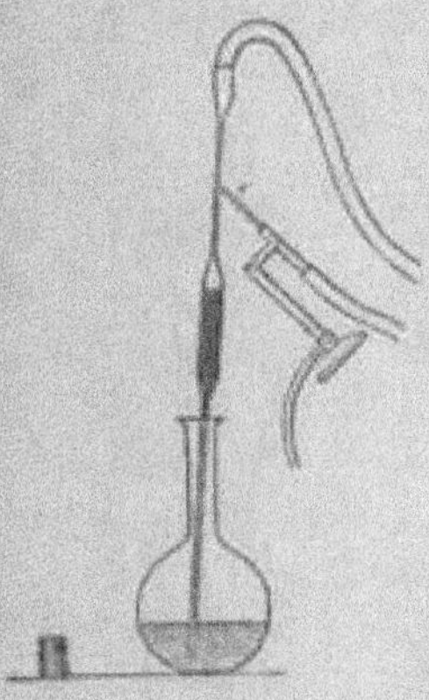

Spontaneous inflammation or cases in which the ignition point is below the ordinary temperature. To demonstrate this phenomenon a tube containing a small quantity of zinc ethyl may be opened and the contents allowed to drop out.

Zinc ethyl may readily be obtained by mixing in a small tubulated retort 50 cubic centimetres of ethyl iodide and 55 grams of dry copper-zinc couple (see Marsh Gas, Exp. 450). The retort is connected by its tubulure to a reflux condenser, the neck being closed with a cork, and is heated in a water-bath until no condensed ethyl iodide is seen to drop back into it. The retort is then removed, the tubulure quickly closed with a cork, and the contents distilled over a rose-burner, a small condenser being used. The distillate may be collected in a small narrow-necked flask. When making this compound on such a small scale, it is not necessary to perform the operations in a current of any inert gas.

The distillate should be transferred to a number of small

glass tubes drawn out and sealed at each end. The tubes are drawn out as shown in figure 114, and by means of a piece of caoutchouc tube the liquid is sucked up into the glass until it is about two-thirds full. The caoutchouc is then closed by a pinch-cock or by being held between the teeth, and the capillary portion above the liquid is sealed by a fine blowpipe jet. The tube is then withdrawn, and the other end closed in the same way.

Fig. 115.

380. A pyrophorus may be opened, and the contents thrown out. These pyrophori are made by reducing suitable salts of certain metals. In the case of iron the best salt to employ is the oxalate; a piece of combustion tube about 25 centimetres long is drawn out at one end, and a quantity of ferric oxalate introduced; the tube is then drawn out at the other end, and a constriction made at a point about 6 centimetres from the end, the whole of the oxalate being in the longer portion of the tube. The salt is then reduced in a current of hydrogen or coal-gas at as low a temperature as possible in a small gas furnace, and as soon as the reduction is complete the tube is removed from the furnace and allowed to cool while the hydrogen is still passing through. When nearly cold the two ends of the tube are sealed up and a small quantity of the black powder shaken into the short portion of the tube beyond the constriction, and the tube is then sealed across at that point. The small tube so obtained serves as a test portion; and if upon opening it and throwing out the contents the material will ignite, the longer portion of the tube may be regarded as sure for the experiment. (See Experiment 391.)

381. Spontaneously inflammable gas may be produced by throwing a few fragments of calcium phosphide into water, or, better, into dilute hydrochloric acid. (See Phosphoretted Hydrogen, Exp. 551.)

382. Silicon hydride, or silicuretted hydrogen, may be generated by adding strong hydrochloric acid to magnesium silicide. A small quantity of the coarsely powdered silicide is placed in a small beaker and drenched with strong hydrochloric acid; silicon hydride is at once evolved as a spontaneously inflammable gas. (See Silicon Hydride, Exp. 513.)

383. The ignition point of gaseous phosphoretted hydrogen is below 100°C., so that a jet of the gas may be ignited by boiling water.

A quantity of phosphoretted hydrogen, prepared by acting upon phosphorus by means of alcoholic potash (see Phosphoretted Hydrogen, Experiment 545), is collected in a gas-holder. A jet of the gas is allowed to escape into the air, and on bringing into it a test tube containing boiling water, the gas will be at once ignited.

384. Carbon disulphide vapour ignites at about 120°C. If, therefore, a few cubic centimetres of the liquid be poured into a small porcelain dish, and a glass rod which

FIG. 116.

has been heated for a few moments in a flame be brought into the vapour, the latter will be instantly ignited.

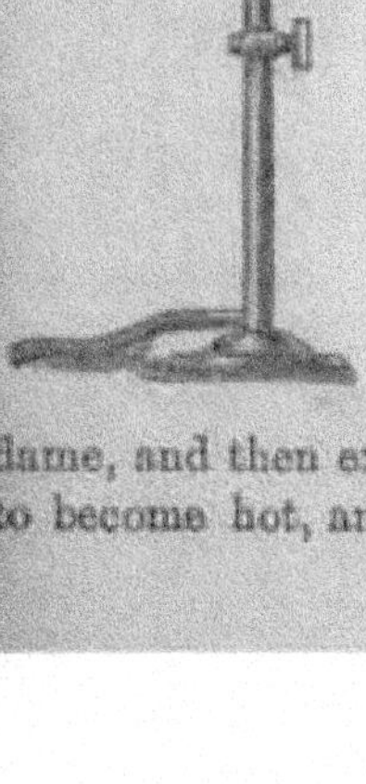

385. The difference between the igniting points of hydrogen and coal-gas may be illustrated by exposing a small bundle of platinised asbestos to a stream of, first, coal-gas and, second, hydrogen. This is best shown by supporting the platinised asbestos at the top of an Argand lamp glass, as shown in fig. 116, and placing the chimney upon the gallery of a small lamp from which the fish-tail or other burner has been removed. The bundle of asbestos is first warmed in a flame, and then exposed to a stream of coal-gas. It will be seen to become hot, and will continue to glow in the coal-gas, but

N

will not cause its ignition.   On substituting hydrogen for coal-gas, it will be seen that as soon as the platinum begins to glow the hydrogen will be ignited.

386. A comparison between the igniting points of hydrogen and marsh-gas may be made by means of sparks emitted by flint and steel. Sparks struck from a flint and steel are made to fall into a stream of hydrogen, as it issues from a fish-tail burner. The hydrogen will be almost instantly ignited, while if similar sparks are caused to fall into marsh-gas they fail to cause its ignition. A shower of sparks is best obtained by

FIG. 117.

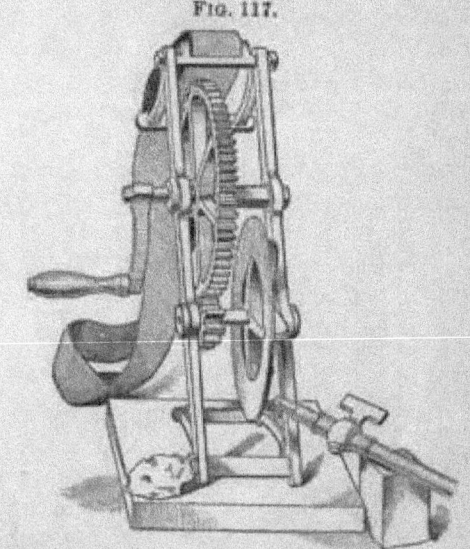

causing a revolving disc of steel to play against a piece of flint pressed lightly against its edge, as in the old ‘steel mill’ of the miners. Fig. 117 shows such a contrivance, in which the disc is made to revolve at a high speed, by means of the tooth gear. It is quite possible, although not quite so easy, to perform this experiment by means of a piece of flint and a flat file. The flint is quickly rubbed down the safe edge of a large flat file, held in such a position that the sparks which are struck off shall fall into the jet of escaping gas.

387. If a piece of copper gauze be depressed upon a Bunsen flame, the metal so rapidly conducts away the heat that the gas which finds its way through the gauze will not be ignited. It may, however, be inflamed by bringing a lighted taper above the gauze, and the actual presence of the gas there be demonstrated. This may be reversed by holding the piece of gauze a short distance above the chimney of a Bunsen before the gas

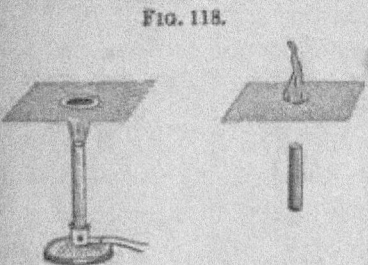

Fig. 118.

is lighted, and then by means of a taper igniting the gas on the upper side of the gauze. The flame will continue to burn upon the gauze, and shows no tendency to strike through. The gas below, therefore, will not inflame, the reason being that the temperature of the metal does not reach the ignition point of the coal-gas. If these experiments be repeated, using a jet of hydrogen in place of coal-gas, the difference between the two gases will be very marked. When the gauze is depressed upon the hydrogen flame, the flame almost immediately passes through the meshes, and it will be found quite impossible to ignite the hydrogen upon the upper side of the gauze without the flame instantly striking back and burning on the under side as well.

CARBON

388. To show the combustion of diamond in oxygen. This is most conveniently done in the apparatus shown in fig. 119. It consists of a perforated ground-glass plate, fitted with a cork, through which pass two stout copper wires. These are con-

nected together at their ends by means of a little strip of
platinum foil, which is creased into the shape of a little trough
or gutter, the ends of the foil being
wrapped round the copper and secured by
a twist of fine binding wire A, fig. 119.
The fragment of diamond is placed in this
platinum trough. The lip of a cylinder is
greased with resin cerate, oxygen gas passed
into it, and the apparatus carrying the
diamond carefully lowered into its place,
care being taken not to allow the diamond
to become lodged against the thick wires
which support the trough. The current from 10 Grove's cells
is passed through the wires, which will heat the platinum to
bright redness, and cause the ignition of the diamond. When
the diamond is once ignited, the passage of the current may be
interrupted, and the diamond will continue its combustion for
some time.

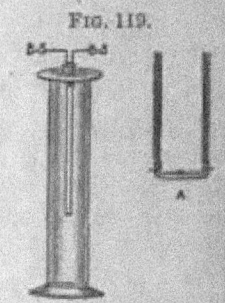

Fig. 119.

A little adjustment is necessary between the size and thickness of the platinum trough, and the strength of current used to
heat the metal. If too much platinum is present, not only does
it necessitate a stronger current to raise its temperature to the
requisite degree, but owing to the rapid conduction of heat the
ignited diamond will not continue its combustion for long without the help of the current. If, on the other hand, the trough
is too slight, there is risk of its being fused by the current.
When the apparatus is first put together, it should be tested
with a current gradually increasing in strength from 5 cells
upwards, and the number of cells that is found necessary to
heat the platinum to the desired temperature should be noted,
conveniently by scratching upon the ground-glass plate. After
the combustion of the diamond has continued for a short time,
the contents of the cylinder are shaken up with clear lime-
water, which will show the presence of carbon dioxide.

389. To show the combustion of graphite in oxygen. A

precisely similar apparatus to the above may be used for this purpose.

390. Combustion of charcoal in oxygen. (See Oxygen.)

391. To show the spontaneous inflammability of finely divided carbon. The pyrophorus is prepared by heating lead tartrate in a glass tube. A piece of combustion tube closed at one end is about half filled with lead tartrate, the tube is then drawn out as shown in fig. 120. By taping the tube horizontally upon the table the salt will spread itself evenly along, leaving a clear space between it and the upper side of the tube. The tube is then heated, either in a short gas furnace, or by means of a Bunsen, in the latter case beginning at the closed end and working along; the heating being continued until vapours

Fig. 120.

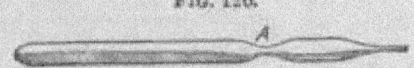

cease issuing from the point. The tube is allowed to cool until it can just be handled, when the point is sealed by a blowpipe, and it is then allowed to get quite cold, without the contents being shaken about. When cold some of the contents are shaken into the part beyond the constriction, and the tube sealed off at A. The small tube serves as a test piece, and if, on filing the end off and throwing the contents out, the particles ignite as they fall through the air, it may be presumed that the main quantity in the longer piece will do the same.

In shaking out the contents of such a pyrophorus, the powder should be allowed to fall through a considerable space of air, the tube being held as high as the operator can conveniently reach, and the powder allowed to fall to the ground.

392. To show the absorption of ammonia by charcoal.[1] A cylinder of ammonia gas is collected over mercury, and one or two fragments of charcoal, about the size of a walnut, which have been heated in a Bunsen flame, are passed up through the

[1] See Tables XXVII. and XXVIII. in the Appendix.

mercury into the gas; in the course of a short time the whole
of the gas will be absorbed by the charcoal.   (See Liquefaction
of Gases, Exp. 645.)

393. To show the absorption of bromine vapour by charcoal.
If a few fragments of charcoal which have been allowed to
absorb bromine vapour are gently heated in a sealed glass tube
of rather large bore, the absorbed bromine is expelled, and is at
once manifest by its colour.   On cooling, the bromine is reabsorbed.   In order to prepare this experiment so that the
same tube may be repeatedly used, it is necessary to get rid of
the hydrogen from the charcoal, otherwise the bromine will
gradually combine with it, and only colourless hydrogen bromide
will be expelled when the tube is next heated.

For this purpose the charcoal, in small fragments, must first
be heated to redness in a combustion tube, and bromine vapour
passed over it (by gently warming a little bromine in a flask
attached to the tube) for some time.   It is then allowed to cool
in the vapour, and afterwards transferred to the tube in which
it is intended to seal it.

394. To show the absorption of sulphuretted hydrogen by
charcoal.   A piece of combustion tube about one metre long is
filled with charcoal, broken into small pieces, about the size of
peas, and a mixture of sulphuretted hydrogen and air, contained
in a small gas-holder, is passed through the tube, in the cold.
The gaseous mixture, which may contain from 10 to 15 per
cent. of sulphuretted hydrogen, is first led through a small
wash-bottle of water to indicate the rate at which it is passing.
If the rate be properly regulated, the gas which issues from the
end of the tube will be found to be perfectly free from sulphuretted hydrogen.   This may be shown by allowing the
issuing gas to bubble through a solution of a lead salt, which
will be seen to remain clear.   In order to demonstrate that the
gas in the holder contains sulphuretted hydrogen, a branch tube
may be introduced between the wash-bottle and the tube con-

taining the charcoal (B, fig. 121), and a few bubbles of the gas allowed to pass direct into a solution of a lead salt, which will be instantly blackened. If a small quantity of mercury be placed in the cylinder C, and the depth of the lead solution be a little less than that in the test tube T, the gas may be made to bubble through C or T at will, by raising or lowering C.

FIG. 121.

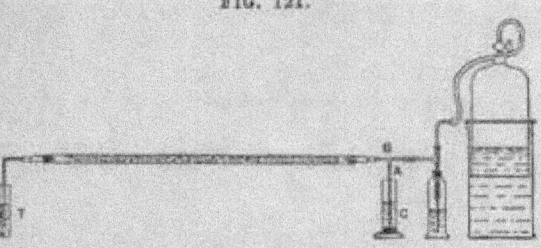

395. To show the evolution of heat by the combination of oxygen with sulphuretted hydrogen, induced by the action of charcoal. This may be done by a slight modification of the above arrangement. A glass tube, about 30 centimetres long and 2½ centimetres wide, is fitted with a cork at each end. Into one cork is fitted a tube upon which is blown a thin tube about

FIG. 122.

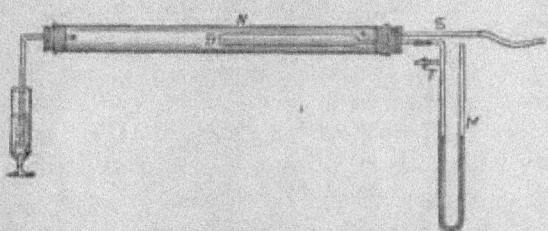

10 centimetres long, and closed at one end, which serves as the bulb of an air thermometer (B, fig. 122); the same cork also carries a short straight tube through which gases are introduced. The cork with these tubes is fitted into the wide tube, which is then filled with charcoal in coarse grains from the other end, the charcoal filling the annular space between the long bulb and the outer tube; when the tube is full the other cork carrying

an exit tube bent at right angles is inserted. The tube from
the bulb is connected with the manometer M, which should be
made with a lateral tube T, by means of which the level of the
coloured water in the two limbs of the tube may be adjusted. A
mixture of two volumes of sulphuretted hydrogen and three
volumes of oxygen is passed from a gasholder through a small
wash-bottle into the tube S, and the escaping gas may be
bubbled through a solution of a lead salt contained in the test
glass. The temperature of the bulb B will rapidly rise, as seen
by the alteration of the level of the liquid in the manometer.

396. A still more striking way of showing this phenomenon
is the following:—

About 5 or 10 grams of powdered charcoal are placed in a
bulb which is blown in the middle of a piece of combustion
tube. A gentle stream of coal-gas is passed over the charcoal,
which is heated by means of a Bunsen lamp until it is perfectly
dry. This point can be ascertained by allowing the issuing gas
to impinge upon a small piece of mirror, and when no deposition
of moisture takes place the heating may be stopped. The
charcoal is then allowed to cool in the stream of coal-gas until
its temperature has fallen so far that the bulb can just be
grasped by the hand, when the coal-gas is replaced by a stream
of sulphuretted hydrogen. The sulphuretted hydrogen should
be allowed to pass over the charcoal for not less than fifteen or
twenty minutes, by which time the bulb and its contents will be
perfectly cold, and the charcoal will have saturated itself with
the gas. (In practice it will be found convenient to prepare the
experiment to this stage, and allow a very slow stream of sulphuretted hydrogen to continue passing through the apparatus
until the final operation is to be made.) The supply of sulphuretted hydrogen is then cut off and a stream of oxygen
passed through ; the tube almost immediately becomes hot, and
moisture is deposited on the glass. The supply of oxygen should
be sufficiently brisk to carry this moisture forward from the
charcoal. In a few moments the temperature rises to the

ignition point of the charcoal, when it bursts into flame and continues burning in the supply of oxygen.

397. To show the withdrawal of gases absorbed by charcoal. A few lumps of charcoal, weighted with lead wire or other convenient material, are thrown into a vessel of water which is placed under the receiver of an air-pump; on exhausting the air, bubbles of gas appear upon the surface of the charcoal, and rise in quantities through the water.

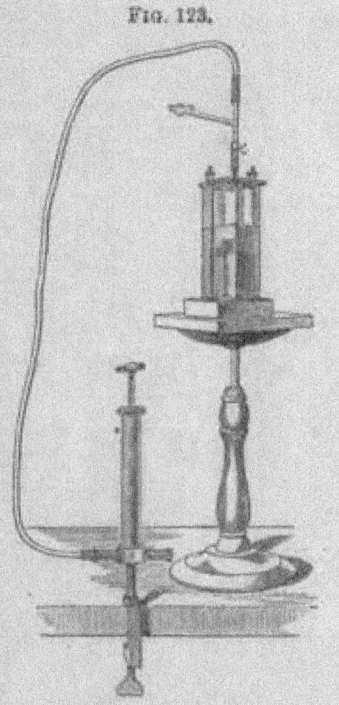

FIG. 123.

This experiment may be shown upon the screen in the following way:—

A square glass cell about 13 centimetres long and 5 centimetres square, and having its lip ground, is fitted with a tight-fitting cover. This is conveniently constructed as shown in fig. 123. A flat brass plate, having india-rubber cemented on the under side, is clamped firmly down upon the mouth of the cell by means of two screws. The plate has a small brass tube soldered into it. One or two small fragments of charcoal are tethered to a little lead weight by threads of cotton, so that they are suspended a little below the surface of the water in the cell. The cell is connected by means of fine 'compo' pipe to a small exhausting syringe, and each stroke of the pump will be followed by the evolution of a number of bubbles from the surface of the pieces of charcoal, which will be very apparent upon the screen.

398. To show that charcoal is specifically heavier than water. This may be demonstrated with the apparatus arranged as in

the above experiment. If, after exhausting, the air be allowed to re-enter, and this operation be repeated once or twice, the charcoal will sink to the bottom.

Pieces of charcoal which have been weighted and thrown for a few minutes into boiling water, will have so much of the included air removed as to cause them to sink when placed in cold water.

399. To show the decolourising power of charcoal. Animal charcoal in fine powder is best suited for this purpose. About 20 grams of the charcoal is introduced into a stoppered cylinder containing water, which is tinted with an aniline colour, and the mixture shaken together. The contents of the cylinder are then thrown upon a filter, and the filtrate will be seen to be entirely colourless.

CARBON DIOXIDE. CO_2

400. To show the presence of carbon dioxide in the air. A quantity of clear lime-water, or baryta water, may be placed in a shallow glass dish, and freely exposed to the air of the room. In the course of half an hour it will appear distinctly turbid.

Fig. 124.

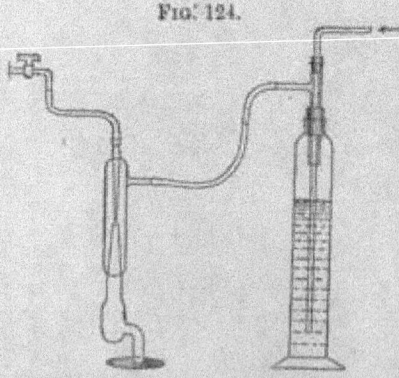

401. Air may be aspirated through baryta water in a Woulf's bottle, or better in a narrow cylinder bottle, by means of a Bunsen's pump (fig. 124); in a few minutes the carbon dioxide will make itself apparent by the turbidity of the solution.

402. To show the carbon dioxide which has been absorbed from the air by mortar. Fragments of old mortar are placed in a test glass and treated with dilute hydrochloric acid :

an effervescence will be observed, and the gas may be shown to be carbon dioxide by any of the usual tests.

403. **To show carbon dioxide in the breath.** A stream of air from the lungs is bubbled through a quantity of clear lime-water, the expiration being continued as long as possible in order to deliver those portions of the breath which are richest in carbon dioxide.

404. The above experiment may be modified in order to show that the effect is not due to the carbon dioxide already contained in the air. Two Woulf's bottles are fitted up as shown in fig. 125, and lime-water placed in each. By means of the mouth-piece E, the air drawn into the lungs will first bubble through the bottle B. When one long inhalation has been made, the air is expired, and it will be compelled to bubble through bottle A. On repeating this process two or three times, the lime-water in A will become turbid, while that in B remains clear.

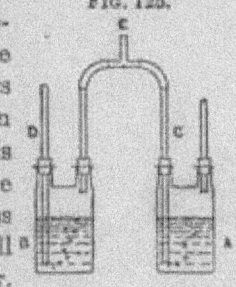

FIG. 125.

The small quantity of breath which passes back into B at each expiration, until the pressure necessary to overcome the column of liquid in A is obtained, is so insignificant that it produces no effect upon the solution in that bottle.

405. **To show the presence of carbon dioxide in natural aerated waters.** A quantity of Apollinaris water is placed in a flask provided with a cork and delivery tube, and gently warmed. Carbon dioxide is evolved and may be passed into lime-water. The natural water may be distinguished from that which is artificially manufactured by not effervescing to any extent when the cork is removed from the bottle, the gas being only very slowly evolved until the water is warmed.

406. **To show the carbon dioxide produced by the combustion of a taper, a candle, or from a gas flame.** A cylinder is held for a second or two over the flame, in order to receive some of the products of the combustion : it is then covered with a glass plate and clear lime-water introduced.

The candle or gas flame may be lowered down into a cylinder, covered loosely with a glass plate, and allowed to continue its combustion for a few moments ; on withdrawing the burning substance and adding lime-water, the carbon dioxide will be evident.

407. To prepare carbon dioxide from marble. $CaCO_3 + 2HCl = CaCl_2 + H_2O + CO_2.$ Fragments of marble are placed in a Woulf's bottle provided with a thistle funnel and a delivery tube ; water is added and strong hydrochloric acid introduced from time to time by means of the thistle tube. The gas may be washed by passing through a Woulf's bottle containing water, and collected at the pneumatic trough.

408. To show the heaviness of carbon dioxide. The gas may be collected by downward displacement of air.

409. The gas may be poured from one vessel into another, and its presence shown in the receiving vessel by means of lime-water.

410. A large tubulated bottle, into the side tubulure of which

FIG. 126.   FIG. 127.

is fitted a wide bent tube (fig. 126) closed with a cork, is filled with carbon dioxide, and the mouth closed with a bung. By first removing the bung and then withdrawing the small cork, the gas will flow out, and may be received in a beaker or cylinder as desired, and its presence shown either by lime-water or by a taper.

411. A large bell-jar may be filled with the gas by displacement. If a beaker suspended by a string be lowered into the gas, and again drawn up, it may be brought out full of the gas, as water is drawn from a well by a bucket.

412. The gas may be siphoned from the bell-jar by an ordinary siphon. The siphon may be started by gently sucking the longer limb, or by first filling it with carbon dioxide, placing the thumb over the end of the long limb and quickly inverting it in the vessel.

413. A large beaker is suspended from the beam of a balance, by means of thin copper wire, and counterpoised. A beaker containing carbon dioxide, which may be filled either from the vessel in Experiment 410 or 411, is gently emptied into the suspended beaker, and the weight of the gas will be evident by the disturbance of the equilibrium of the balance.

414. The vessel described in Experiment 410 may be placed a few feet from a screen, and a strong shadow of it cast by means of the electric light (the naked light without lenses). When the corks are removed the stream of gas as it flows out at the lower tube will be seen upon the screen as a descending current, owing to the difference between the refractive power of carbon dioxide and air. If a beaker be placed to receive it, the gas will be visible as it overflows.

A similar effect is seen, and more visibly in a large room, by pouring out in front of the screen the contents of a large bell-jar filled with the gas.

415. To float soap bubbles upon carbon dioxide. A large bell-jar is filled with the gas, as in Experiment 411. For this experiment, however, the gas must be washed by being bubbled through water, to remove any hydrochloric acid that is carried over with it, the presence of this gas being fatal to the life of a soap-bubble.

The bubbles are best blown upon a piece of glass tube about 12 centimetres long and 7 millimetres bore. This tube is partially choked by pushing a piece of cork into it, the cork having a small notch cut out of it. A bubble of about 8 or 9 centimetres diameter is blown and the tube held in such a way that the top of it may be covered at will by the finger (fig. 128a).

While conveying the tube with the bubble to the mouth of the bell-jar, the tube should be uncovered; when it is wished to drop the bubble, the finger is placed upon the top of the tube and a slight jerk given by the wrist. If the bubble be allowed to fall

FIG. 128.

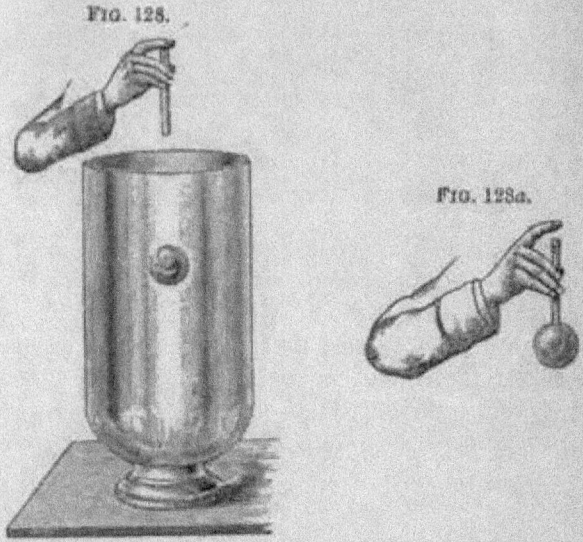

FIG. 128a.

vertically into the bell-jar, it will remain for some considerable time floating upon the gas. (For preparation of the soap solution see Hydrogen, Experiment 14.)

416. The heaviness of carbon dioxide, and its behaviour towards ordinary combustibles, may be simultaneously demonstrated by pouring the gas from a cylinder, or beaker, upon a burning taper or candle. The candle is stood in the open, and the gas contained in a large beaker is quickly poured over it from a height of 10 to 20 centimetres.

The candle may be placed in a cylinder, and the gas poured in upon it. In this case the cylinder should not be too narrow, or the up current of hot air from the candle will prevent the carbon dioxide from entering the mouth.

417. A number of candles, conveniently 3 or 4, may be arranged on a light stand, by means of pieces of wire soldered to the upright support at varying heights. The end of each wire is bent into a helix, and a short piece of candle fastened to it by slightly warming the wire. The candles are lighted and the apparatus lowered into a large bell-jar, and a stream of carbon dioxide delivered down to the bottom of the vessel by a piece of flexible tube. As the vessel gradually fills with the gas, the candles will be extinguished in order from the bottom.

418. A quantity of benzene (about 50 cubic centimetres) is placed in a porcelain dish, or a soup plate, and inflamed. A

FIG. 129.

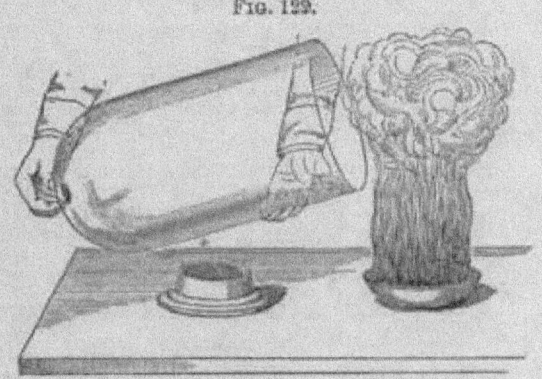

large bell-jar full of carbon dioxide is emptied over it, when the conflagration is at once extinguished. The gas should not be poured down immediately upon the top of the flames, but rather to one side, as shown in the figure.

419. To show the combustion of potassium in carbon dioxide $2K_2 + 3CO_2 = 2K_2CO_3 + C.$ A stream of the gas from a generating apparatus is washed through water, and passed through a bulb tube containing a fragment of potassium. The metal is heated strongly with a Bunsen lamp, and will inflame in the carbon dioxide, leaving a black deposit containing carbon upon the glass.

The most convenient form of apparatus to employ, when a regular stream of carbon dioxide is required is seen in fig. 130.

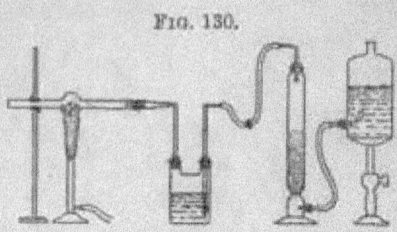

FIG. 130.

The eprivette is filled with fragments of marble which are prevented from dropping through the constriction by first placing in the cylinder a roughly rounded piece of pumice - stone, sufficiently large to block the passage.

A mixture of hydrochloric acid and water in about equal volumes is introduced into the tubulated bottle, which is placed upon a stand, or upon blocks, of such a height that the acid cannot rise in the eprivette more than about two-thirds of the height.

420. To show the combustion of magnesium in carbon dioxide. A strip of magnesium ribbon is ignited and lowered into a cylinder of carbon dioxide. The metal continues its combustion in that gas, and on rinsing out the contents of the vessel, and throwing them upon a filter, black particles containing carbon will be seen in considerable quantity.

421. To show the action of carbon dioxide upon litmus solution. A stream of washed gas is made to bubble through a quantity of litmus solution contained in a porcelain evaporating dish, or a white soup plate. The tube conveying the gas may be drawn to a point, and the liquid gently stirred with the tube. In a few moments the colour of the litmus will be changed from blue to a claret red. This reddened solution may now be divided into two portions. To one a drop of a strong acid, e.g. hydrochloric acid, should be added, to show the difference between the red so obtained and that produced by carbon dioxide, while the other may be heated to show that the original blue of the litmus will be restored.

422. To show the absorption of carbon dioxide by sodium hydroxide. A glass tube from ¾ to 1 metre long, upon one end of which a dropping funnel has been blown, is filled with carbon dioxide over water. A quantity of a moderately strong solution of sodium hydroxide is placed in the globe of the funnel, and allowed slowly to enter through the stopcock. As the solution trickles down the sides of the long tube, it absorbs the gas, and the water gradually rises in the tube until it is completely full.

423. To show the effect of pressure upon the solubility of carbon dioxide in water. Two rather wide glass tubes, about one metre long, and as nearly as possible of a uniform bore throughout, are closed at one end and connected by means of a bridle tube to the same mercury reservoir (fig. 53). The tubes are connected to the bridle tube by means of caoutchouc stoppers, and in order that these may be wired securely into place, the edges of the wide tubes should be slightly opened out in the form of a lip. One of these caoutchouc stoppers carries a short stopcock, in order that the level of the liquid in the two tubes may be finally adjusted. Both the tubes are filled with carbon dioxide, one being dry, while the other contains a small quantity of water, occupying about 3 centimetres of the tube, which has been previously saturated with carbon dioxide. The height of the mercury in the dry tube should be made level with the top of the short column of water in the other, and the position marked by india-rubber bands upon the tubes. Two other bands are placed on the tubes at a point half-way between the bottom rings and the top. On raising the reservoir until the gas is subjected to a pressure of an extra atmosphere, it will be seen that the volume of gas in the tube containing the water is diminished more than that in the other. If a short time be allowed for solution to take place, it will be seen that the water will take up its own volume of the gas under the increased pressure of one atmosphere, and the mercury in each tube will stand at the same level. On lowering the reservoir, and allowing a little time to elapse, the gas dissolved under the increased pressure will be again evolved, and the mercury in the one tube will be level with the top of the column of water in the other.

o

To show the liquefaction of carbon dioxide by pressure. (See Liquefaction of Gases.)

424. To collect solid carbon dioxide. This is effected by causing a stream of the liquefied gas to issue from an iron bottle through a fine jet, into a metal box specially constructed for the purpose. Fig. 131 shows the usual form of the apparatus.

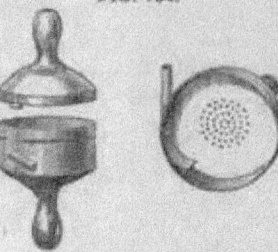

FIG. 131.

The side tube receives the nozzle of the gas bottle, and the issuing stream of liquid is made to impinge upon the curved tongue of metal, and is thereby caused to revolve inside the box. The escaping gas issues by the perforated plates through the insulated wooden handles. As the solid collects in the box, small fragments are driven out through the handles, and the operator can tell by the quantity so escaping when the box is full. As the ordinary pressure gauge is not available for ascertaining what quantity of carbon dioxide is contained in the cylinder, it is well to know the weight of the bottle when free of liquid, and also its weight when full. To avoid mistakes, these weights should be made with the nozzle and handle removed. The solid when collected should be removed from the metal receiver and placed in a cardboard box, which may be thickly covered with felt.

The substance should be handled with some care, a wooden spatula or paper knife being the best for cutting and lifting it. It may be placed upon the hand or even upon the tongue with impunity, but if lifted with the fingers care must be taken to avoid squeezing it, as, the pressure bringing it in contact with the skin, the intense cold will produce unpleasant blisters.

425. To show the evolution of carbon dioxide gas from the solid. A fragment of the solid may be placed in an empty soda-water bottle, which is then gently corked up; in a few moments the cork will be forcibly ejected.

426. A fragment may be placed in a small flask fitted with a cork and delivery tube, the latter dipping into lime-water. The lime-water will be at once rendered turbid.

427. A fragment placed in a cylinder, loosely covered with a glass plate, will rapidly displace all the air, and fill the cylinder with the gas, which can be tested by a taper or by lime-water.

428. A fragment may be placed in a flask with delivery tube, and the gas collected at the pneumatic trough, when the regular and steady evolution of the gas will be seen.

429. A small fragment may be placed upon the tongue, and the evolving gas gently breathed upon a lighted candle or taper, which will be at once extinguished.

430. To freeze water with solid carbon dioxide. A small tin can with a wire handle is placed upon a wooden stand, upon which a small quantity of water has been poured; a few small pieces of the solid are thrown into the tin, and moistened with a little ether. The water will be almost instantly frozen, the solidification being accompanied by a crackling sound, and on lifting the can by its handle it will be seen that it is securely cemented to the stand.

431. To freeze mercury by means of solid carbon dioxide. About 80 cubic centimetres of mercury are placed in a round-bottomed porcelain basin.

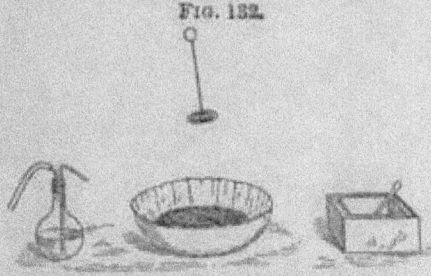

FIG. 132.

In order that the mercury, when solidified, may be lifted from the dish, a stout copper-wire, bent into a flattened helix, is held in the mercury while it is being frozen; and, in order to prevent the frozen mercury from adhering to the dish itself, the latter is lined with paper, by pressing into it, with the fist, a piece of writing paper, the edges of which may be trimmed with scissors A quantity of solid carbon dioxide is placed upon the mercury, and the copper-wire is then introduced. A small quantity of ether is then poured upon the mercury, and in a few moments a solid crust begins to form, and the wire will be held in its position by the frozen mercury. More solid carbon dioxide may be added until the mercury is entirely solidified throughout. The ether for these experiments should be fairly free from water and alcohol, and is conveniently used in a small wash-bottle.

432. The malleability of the solid mercury obtained above may be shown by lifting the lump from the dish and placing it upon a block of iron, which is wrapped in felt or thick flannel. It should be struck with a hammer, upon the head of which a piece of thick leather has been cemented. With such a hammer and anvil the mercury may be beaten into any shape, like a lump of lead, and little or none of it will be melted in the operation.

433. The solid mercury, after being hammered as above, may be plunged into water, by doing which the mercury will be

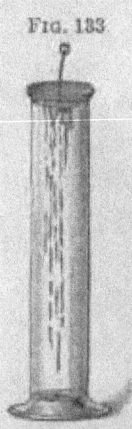

Fig. 133

melted and the water frozen. This is strikingly shown by filling a tall, wide glass cylinder with water, and placing upon the top of the cylinder a piece of wire gauze, bent into the shape of a cup or dish. On placing the solid mercury into the water contained in this gauze dish, the metal melts, and running through the meshes, forms icicles several centimetres in length hanging from the gauze, and leaves a solid piece of ice in the cup adhering to the copper wire, by which it can be bodily withdrawn (fig. 133). It is well to use water which is only a few degrees above the freezing-point, and to throw a ray of strong light upon the vessel while the experiment is being performed.

434. The composition of carbon dioxide may be demonstrated by burning a small fragment of charcoal in a measured volume

of oxygen. A small globe or flask, blown upon a U tube, is fitted with a stopper through which pass two thick wires. One of these wires carries a small bone-ash crucible made by nearly filling a deflagrating spoon with bone-ash. The second wire ends just above the crucible, and is connected by a small loop of fine platinum wire to the other thick wire, just above the crucible (fig. 134).

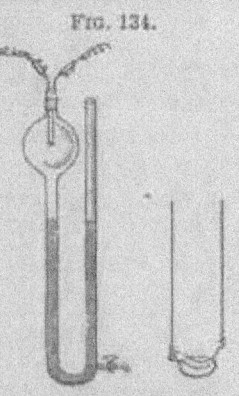

Fig. 134.

Mercury is placed in the U tube, and its height indicated by a band upon the tube, and the globe is filled with oxygen by displacement. A small fragment of charcoal is then placed upon the crucible in such a position that the loop of platinum wire is touching it, and the stopper carefully inserted. By means of an electric current the platinum wire is heated, and the little piece of charcoal ignited, which continues burning until it is entirely consumed. At the conclusion of the experiment, and when the apparatus is again cold, the mercury will return to the position that it originally occupied.

The charcoal for this experiment must be free from hydrogen. This is most easily secured by heating a quantity of charcoal to bright redness in a stream of chlorine, the charcoal being in pieces about the size of a pea, and contained in a piece of combustion tube.

CARBON MONOXIDE.

PREPARATION.

435. By the action of sulphuric acid upon a formate. $H.COOH = H_2O + CO$. About 10 grams of barium formate are placed in a small flask fitted with a dropping funnel and a delivery tube. Strong sulphuric acid is introduced in small

quantities at a time by means of the funnel, and a rapid evolution of carbon monoxide at once takes place. The gas, which is almost perfectly pure, may be collected over water in the ordinary way.

436. By the action of sulphuric acid upon oxalic acid $C_2H_2O_4 = H_2O + CO_2 + CO$. The gas is generated in a small flask fitted with a mercury safety funnel and a delivery tube. As there is an equal volume of carbon dioxide produced in this reaction, it is well to pass the gas through a series of Woulf's bottles. The first one should contain baryta water to demonstrate the presence of the dioxide; and if the gas be then passed through two others containing a solution of caustic soda, the monoxide can be collected sufficiently free from the dioxide for all practical purposes. Baryta water is preferable to lime-water for the first bottle, as with the latter the precipitated carbonate is so rapidly redissolved by the excess of carbon dioxide that after the first few minutes the liquid gives no visible evidence of the presence of that gas. 20 or 30 grams of crystallised oxalic acid are put into the generating flask, and covered with strong sulphuric acid; on the application of a gentle heat the gas is readily evolved, and its rate of evolution may be controlled to a nicety by a little regulation of the lamp.

437. By the action of sulphuric acid upon potassium ferrocyanide. $K_4FeC_6N_6 + 6H_2SO_4 + 6H_2O = 2K_2SO_4 + FeSO_4 + 3(NH_4)_2SO_4 + 6CO$. About 30 grams of crystallised ferrocyanide is roughly broken up and put into a flask, and 300 cubic centimetres of strong sulphuric acid added. The flask, which should be of such a size that the materials used do not more than one-third fill it, is fitted with a safety funnel and a delivery tube. As the gas is usually contaminated with small quantities of carbon dioxide, it is well to pass it through a Woulf's bottle containing caustic soda solution. The temperature at which the decomposition takes place being rather high, it takes a considerable time to heat the mixture to the required degree. If the heat be applied rapidly, by the use of a large flame, there is great liability of

the gas being suddenly evolved with great energy, the contents of the flask frothing up, and the reaction becoming altogether uncontrollable; to avoid this it is well to heat the flask with a moderate rose-burner, and carefully regulate the flame when the first signs of effervescence appear.

438. By the action of red-hot charcoal upon carbon dioxide. $CO_2 + C = 2CO$. A piece of iron gas-pipe is fitted at each end with a cork carrying a short straight glass tube. The iron pipe is filled with small fragments of charcoal, and is heated either by means of a coke furnace or an Erlenmeyer's gas furnace to a bright-red heat. On passing a gentle stream of carbon dioxide, generated in a constant apparatus (see fig. 130), through the hot tube, the issuing gas may be inflamed at the glass exit tube, where it burns with the characteristic blue flame of carbon monoxide.

439. To show that the gas evolved by the action of sulphuric acid upon oxalic acid is composed of the two oxides in equal volumes. A long tube (see Carbon Dioxide, Exp. 422) with a dropping funnel upon one end, is filled with the gases as they are evolved from the mixture, by collection over water. The tube is divided into two equal parts by an india-rubber ring, and on allowing a solution of sodium hydroxide to enter, the carbon dioxide will be absorbed, the water in the dish rising to the graduation. The tube may then be removed from the dish, the mouth being closed with the thumb, and inverted, and the carbon monoxide ignited. In case the tube is not of uniform bore throughout, and also to allow for the diminution of pressure caused by the rising column of water in the tube as the dioxide is absorbed, it is well to graduate the tube by a trial experiment. The tube is filled with the mixed gases from the oxalic acid, and the carbon dioxide absorbed by sodium hydroxide; the ring is then fixed at the point to which the liquid rises in the tube.

440. The combustion of carbon monoxide and its charac-

teristic flame may be illustrated by igniting a cylinder of the gas, collected by any of the above methods for its preparation, or better by burning the gas as it issues from a jet; in this case the jet should be a wide piece of glass tube, and the gas should be contained in a gas-holder.

441. The behaviour of carbon monoxide towards ordinary combustibles may be demonstrated by plunging a lighted taper into a cylinder of the gas, when the gas will be inflamed but will not support the combustion of the taper.

442. The formation of carbon dioxide by the combustion of carbon monoxide may be shown by holding a clean cylinder for a moment over a burning jet of the gas, and then pouring into the cylinder some lime-water.

This is more strikingly done, and at the same time the fact demonstrated that carbon monoxide is without action upon lime-water, by pouring into a cylinder of the gas some clear lime-water and agitating the solution, which will be seen to remain clear. On applying a light to the gas, and again shaking the contents of the vessel, abundant evidence of carbon dioxide will be seen.

For this experiment the gas must be carefully purified from the dioxide. It should be collected in a stoppered bottle, and a small quantity of caustic soda solution introduced and well agitated with the gas. The gas should then be decanted into a stoppered cylinder, the stopper of which is greased with resin cerate, in which it may be preserved for any length of time.

443. To explode a mixture of carbon monoxide and oxygen. Two volumes of the monoxide and one of oxygen are introduced into a soda-water bottle, and a light applied to the mouth, when the mixture explodes with some violence.

444. To show that a perfectly dry mixture of carbon monoxide and oxygen will not explode. A straight eudiometer of rather wide bore glass tube is made, and two platinum wires sealed in at a short distance from the closed end. The platinum wires should be rather thick, and the ends which are

to be inside the tube should be fused in an oxyhydrogen flame until there is a little ball or knob of platinum formed; the object being to avoid heating the platinum by the passage of the spark more than is absolutely necessary.

To prepare the experiment the tube is first carefully heated all over, to remove as far as possible the moisture from its surface, while still hot it is filled with mercury which has been heated nearly to its boiling-point, and allowed to so far cool down that it is possible to manipulate it. A mixture of the two gases, in the proper proportions, is made to pass through a series of drying tubes containing pumice moistened with sulphuric acid, and finally through a tube filled with phosphorus pentoxide, and is then allowed slowly to bubble up into the eudiometer, until about 15 centimetres of the tube are occupied with gas. A cylinder of phosphorus pentoxide is then introduced up the tube, through the mercury. This solid cylinder of the pentoxide is easily and quickly made by taking a short straight glass tube of the required bore, closing one end with a cork, and filling it with the anhydride, which is then firmly pressed down upon the closed end with a piston made by fastening a cork upon a piece of glass rod. On withdrawing the cork from the end of the tube the pentoxide may be pushed out by the piston as a compact little cylinder; it should be pushed out on to the mercury in the trough and instantly depressed beneath the surface, and put up into the eudiometer. The eudiometer with its contents should then be removed from the trough and stood in a short glass cylinder containing mercury; a layer of glycerine should be poured upon the mercury in the little cylinder, to prevent any passage of air down between the metal and the glass, and so into the eudiometer. The whole must then be allowed to stand for several weeks—if possible, for six months—to allow of the complete removal of the moisture.

FIG. 135.

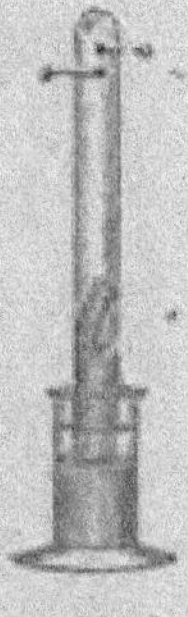

When it is intended to show that the mixture will not explode by the passage of the electric spark, it is necessary to

use only a rather feeble spark, and to allow only one at a time
to pass; the hammer of the induction coil being held in the
fingers, so that the passage of the sparks may be regulated at
will. It will be seen that, as each spark passes, a feeble combustion takes place in its immediate path, producing a little
halo of light, in colour resembling the flame of burning carbon
monoxide, but the combustion will not be propagated, and the
contents of the tube will not explode. If the spark be passed
before sufficient time has been allowed for the gases to become
perfectly dry the combustion may be propagated, but not with
explosive violence, the flame merely travelling slowly down the
tube. When once prepared, the tube can be preserved for any
length of time.

445. To show the action of chlorine upon carbon monoxide
under the influence of light. The formation of carbonic oxydi-
chloride, or 'phosgene gas,' $COCl_2$. Two equal-sized cylinders
are filled, one with chlorine and the other with carbon monoxide,
both gases being collected by displacement. The two vessels
are placed mouth to mouth and the gases allowed to mix, and
the cylinders again covered with glass plates. The contents of
both cylinders will show a slight yellow colour, from the presence of the chlorine. If one of these be exposed for a short
time to the electric light, the colour will rapidly disappear, and
the contents of the cylinder, when allowed to escape into the
air, will evolve dense fumes, owing to the decomposition of the
oxychloride in contact with the moisture of the air, into hydrochloric acid.

446. To prepare nickel carbonyl, $Ni(CO)_4$. A combustion
tube 70 or 80 centimetres long is filled with nickel oxide, and
the oxide reduced in a stream of hydrogen at a dull-red heat;
the operation of reduction is complete when no deposition of
moisture is seen to take place upon the surface of a piece of
mirror held in the issuing stream of hydrogen. The tube is
then allowed to cool with a slow stream of hydrogen passing
through it. It will be found convenient to fill three such tubes.

for, although the experiment can be performed with one, the yield of the compound is small. If the tubes are corked up they may be preserved if necessary for a day or two before being used, but it is more advantageous to use them as soon after the oxide has been reduced as possible.

The tubes are arranged side by side, and a stream of carbon monoxide, after being bubbled through sulphuric acid, is passed through them. The issuing gas, which consists of carbon monoxide charged with nickel carbonyl, is led through a condensing tube in a freezing mixture, where the nickel compound is condensed as a colourless, mobile, and highly refracting liquid. A certain quantity, however, of the compound escapes condensa-

Fig. 136.

tion, and is carried forward with the excess of carbon monoxide. As the vapour of nickel carbonyl is highly poisonous it is necessary that the escaping gas be either delivered into a draught flue, or, what is better, it may be collected in a second gas-holder and used over again; in this way a much smaller quantity of carbon monoxide will be required. By continuing the operation for a few hours 20 or 30 cubic centimetres of the liquid can readily be obtained, which may be preserved in a sealed tube.

In order to seal it up it is poured from the condensing tube into a glass tube closed at one end, and constricted at a short distance from the other, the tube being placed in a freezing mixture. When the liquid has been introduced, the air is displaced by delivering dry carbon dioxide into the tube by means of a piece of finely drawn-out glass capable of passing the constriction, and the tube then sealed at the constricted part by means of a small blowpipe flame.

447. **To show the character of the flame of burning nickel carbonyl,** the gas which escapes condensation in the freezing mixture may be used. The gas is caused to pass through a piece of hard glass tube and ignited as it issues from the end, when it will be seen to burn with a highly luminous flame, producing a smoke of nickel oxide.

If a white porcelain dish be depressed into the flame a black deposit of nickel is obtained, but if the dish be held near the tip of the flame a greenish deposit of nickel oxide will be formed.

448. **To show the decomposition of nickel carbonyl by heat.** The hard glass tube above mentioned may be heated by a Bunsen flame, when a lustrous mirror of nickel will be deposited upon the glass; at the same time the luminous character of the flame will disappear, until, if the gas be not passing too fast, the flame will assume the blue colour characteristic of burning carbon monoxide.

In this and the former experiment it is advisable to raise the tube in the wash-bottle above the surface of the acid, or to tip the bottle until the gas no longer bubbles through the liquid, in order that the issuing gas may burn without intermission.

METHANE (Marsh Gas), CH_4

449. **Preparation from sodium acetate and soda-lime.** $CH_3.COONa + NaHO = Na_2CO_3 + CH_4.$ A quantity of sodium acetate (about 30 grams) is mixed with three times its weight of soda-lime, and the mixture strongly heated in a copper flask.

The gas may be washed through strong sulphuric acid in a Woulf's bottle to remove any illuminating hydrocarbons with which it may be mixed. If it be only required to perform the experiment on a small scale, a Florence flask may be substituted for the copper one; the flask must be supported in a position nearly horizontal, with its neck directed slightly downwards so that condensed moisture may not run back upon the hot glass.

450. Preparation by the action of zinc-copper couple upon methyl iodide. $CH_3I + CH_3.HO + Zn = ZnI(OCH_3) + CH_4$. The zinc-copper couple for this experiment may be prepared as described under Hydrogen, Experiment 3. A much more convenient and expeditious method of obtaining the couple is simply to add to dry zinc-dust about one-tenth of its weight of dry copper oxide, and shake the two powders together in a flask; the couple so prepared is equally as efficient as that produced by the usual method, and may be employed in the experiment above referred to for the decomposition of water. In cases where the couple is required dry, this method is especially advantageous, as it at once does away with the troublesome process of drying the couple which has been prepared by the wet method. Some of the couple is placed in a small and rather wide-mouthed flask fitted with a cork, through which pass a dropping funnel and a scrubber tube (fig. 137). This latter is a long piece of wide tube, one end of which has been drawn out or fused on to a piece of smaller tube. This tube is filled up with granulated zinc which has been immersed in a dilute solution of copper sulphate, and fitted at the upper end with a cork carrying a delivery tube. A mixture of methyl iodide and methyl alcohol in equal volumes is placed in the dropping funnel and allowed to fall upon the copper couple. The evolved gas as it passes up through the tube containing the

FIG. 137.

coppered zinc, will be scrubbed free from vapour of methyl iodide, and the gas may be collected over water in the usual way.

451. Preparation by the action of water upon zinc methyl.
$Zn(CH_3)_2 + 2H_2O = Zn(HO)_2 + 2CH_4$. A small quantity of zinc methyl is diluted with dry ether until the liquid is no longer spontaneously inflammable. The ether is best dried by throwing into it a few fragments of freshly cut metallic sodium and leaving it to stand for a few hours.

The ethereal solution of zinc methyl is poured into a short glass tube closed at one end; the tube so filled is covered by the finger and the contents decanted up into a cylinder filled with water and inverted in a pneumatic trough. The ethereal solution rises within the cylinder and marsh gas is disengaged with the precipitation of zinc hydrate. The gas should be agitated with the water to wash out of it as much of the ether vapour as possible before applying a light to it.

452. To show the inflammability of marsh gas and the non-luminosity of the flame. The gas for this experiment must be specially purified. The sample obtained by the action of water

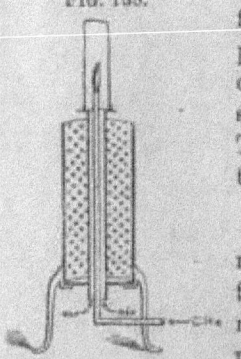

FIG. 138.

upon zinc methyl will burn with a luminous flame owing to the ether vapour which is present, but that produced by either of the other methods described, if washed through sulphuric acid, will answer the purpose. The gas should be burnt from a clean fish-tail burner.

453. That marsh gas when heated to redness will burn with a brightly luminous flame may be shown in the following manner. A piece of narrow iron gas-pipe, surrounded by a piece of wider iron pipe, is made to pass through a charcoal fire contained in a perforated sheet-iron cylinder (fig. 138). In this way both the iron pipes become heated to redness. The smaller inside tube is connected to a gas-holder containing marsh gas, and a small jet of the gas ignited as it issues from the pipe. An ordinary Argand lamp-glass is then placed upon a flange which is fastened to the outer iron tube, so that the air which supplies the flame also becomes heated to redness by its passage up the space between the two pipes. Under these circumstances the luminosity of the burning marsh gas is considerably greater than that of ordinary coal gas.

454. The explosion of a mixture of marsh gas and oxygen may be shown by introducing into a soda-water bottle a mixture of one volume of marsh gas and two of oxygen and applying a light to the mouth of the bottle.

ETHYLENE. C_2H_4

455. **Preparation from alcohol by the action of phosphorus pentoxide.** $C_2H_6O - H_2O = C_2H_4$. A quantity of the anhydride is placed in a small flask and a few drops of absolute alcohol added. A cork carrying a short straight glass tube is inserted and the flask gently heated. Ethylene is rapidly evolved and may be inflamed as it issues from the tube.

456. **Preparation from alcohol by the action of sulphuric acid.** A mixture of 4 volumes strong sulphuric acid and 1 volume water is heated to 165° in a flask capable of containing at least four times the volume of the acid mixture used. The flask is fitted with a cork carrying an exit tube, a thermometer which reaches down into the liquid, and a straight wide tube ('combustion' tube size) extending nearly to the bottom of the flask. Into the top of this tube is fitted a dropping arrangement, such as either A or B, fig. 139, whereby the rate at which the alcohol

is admitted can be seen. The lower extremity of this little arrangement is drawn out and bent as shown in the figure so that the alcohol as it enters shall run down the inside wall of the wide tube and be delivered beneath the surface of the acid. The alcohol is supplied from a small reservoir (conveniently a stoppered funnel) connected to the apparatus by a rubber tube. The chemical reaction is always accompanied by secondary decompositions. resulting in the blackening of the mixture through the separation of carbon; and this carbon then acts upon the sulphuric acid with the evolution of carbon dioxide and sulphur dioxide. It is necessary therefore to pass the gas through one or two Woulf's bottles containing caustic soda solution if the reaction is to be continued for any time. A still more inconvenient result of this charring is that the mixture froths up (hence the necessity for a capacious generating flask) in a most troublesome manner. This may be lessened by adding sand to the mixture, or by placing a thin layer of melted paraffin upon the surface: or the charring and its attendant troubles may be considerably postponed by admitting alcohol vapour instead of liquid alcohol into the acid. In this case the alcohol is boiled in a separate and small flask and the vapour conveyed beneath the acid by a narrow glass tube. The disadvantage of this last plan is the greater difficulty of regulating the admission of the alcohol.

457. Preparation from alcohol and tribasic phosphoric acid. The most convenient method for the preparation of ethylene when any quantity of the gas is required, or when it is wanted in a state of purity, is the following, in which tribasic or 'ortho' phosphoric acid is substituted for sulphuric acid. In this case no charring of the mixture takes place; consequently there is no frothing, and therefore the operation may be carried out in vessels of small dimensions. Also since the gas is entirely free from carbon dioxide and of course contains no sulphur dioxide, all purifying vessels for the removal of these gases are dispensed with.

About 50 or 60 c.c. of syrupy phosphoric acid [1] of sp. gr. 1·75 are placed in a small Wurtz flask of about 180 c.c. capacity. The flask is fitted with a cork carrying a thermometer and a dropping tube, the end of the latter being drawn out to a fine tube and reaching to the bottom of the flask. Phosphoric acid of sp. gr. 1·75 boils at a temperature about 160°. It is heated

FIG. 139.

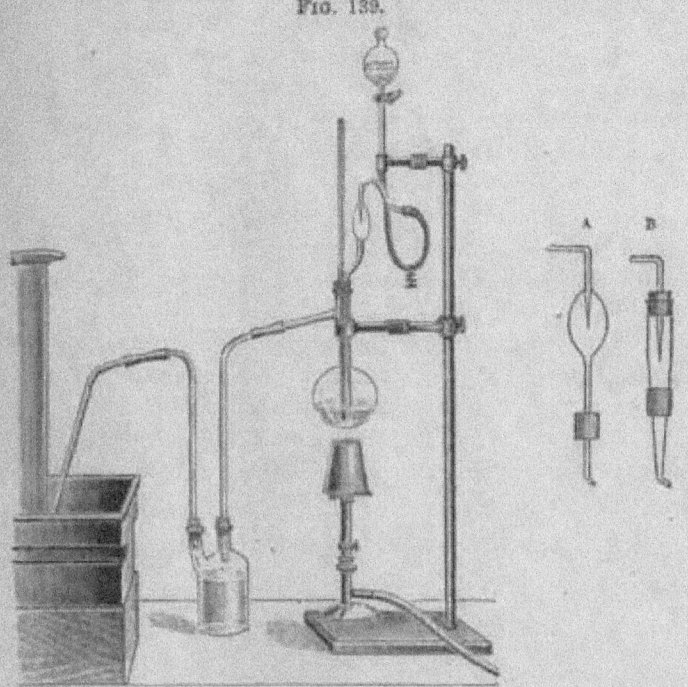

in the flask and allowed to boil for a few minutes until its temperature reaches 200°, when ethyl alcohol is allowed to enter drop by drop by means of the dropping arrangement shown in the figure. Ethylene is immediately disengaged, and by main-

[1] Syrupy phosphoric acid of this strength is an article of commerce, and may be purchased at about 1s. per pound. This acid is the 'ortho' or tribasic phosphoric acid. 'Glacial' acid or 'meta' phosphoric acid will not answer the purpose

P

taining the temperature between 200° and 220° a continuous supply of the gas can be obtained even from so small a generating apparatus, at the rate of 10 to 15 litres per hour, so long as the supply of alcohol is kept up. The gas should be conducted through a small Woulf's bottle (100 to 150 c.c. capacity) standing in a beaker containing a little ice, in which an aqueous liquid collects containing a small quantity of ether, a little alcohol which escapes decomposition, and traces of an oily liquid floating upon the surface. The same quantity of phosphoric acid is capable of decomposing an indefinitely large quantity of alcohol into ethylene and water: the process therefore is *continuous*, and when once set agoing may be continued with scarcely any attention for days together. It is found, however, that after prolonged heating the glass vessel shows signs of being acted on by the phosphoric acid; it is advisable therefore to protect the thermometer by wrapping the extremity of it in a strip of platinum foil.

It is preferable to use pure ethyl alcohol for the preparation, and as the quantity required is so extremely small this is not a serious matter. Methylated spirit may however be used, only in this case the ethylene will be contaminated with more or less methyl ether. This gas being readily soluble in water is, however, easily removed.

458. Preparation by the action of alcoholic potash upon ethyl iodide. $C_2H_5I + KHO = KI + H_2O + C_2H_4$. This may be shown in a precisely similar manner to that described for the preparation of acetylene (see Exp. 466), substituting ethyl iodide for the ethylene dibromide. As this reaction is not very rapid, it is necessary to employ the apparatus with the two reflux condensers, otherwise the alcohol will boil off before any quantity of ethylene is obtained.

459. Preparation by the action of zinc-copper couple upon ethylene dibromide. $C_2H_4Br_2 + Zn = ZnBr_2 + C_2H_4$. Ethylene dibromide is diluted with its own volume of alcohol, and a few cubic centimetres of the mixture poured upon a quantity of zinc-

copper couple contained in a small flask with a wide delivery
tube. The mixture gradually becomes warm, and ethylene is
evolved. The flask may be very gently heated to start the reaction, but the source of heat must be removed as soon as
the action has commenced, as considerable rise of temperature
accompanies the reaction and the evolution of gas is liable to
become .tumultuous; for this reason a wide delivery tube is'
desirable.

460. The combustibility of ethylene, and the luminosity of
the flame may be shown by burning the gas at an ordinary fish-
tail burner, the gas being previously collected in a gas-holder.

461. The absorption of ethylene by bromine may be illustrated
by placing in a round-bottomed flask, or ' bolt head,' a small glass
bulb of bromine, and then filling the vessel with ethylene over
water. The flask is closed by a caoutchouc cork carrying a glass
stopcock. The bulb is broken by shaking against the sides of the
vessel and the bromine exposed to the gas by causing it to run
over the sides of the vessel. It will be seen that the colour of the
bromine disappears, and an almost colourless oily liquid is formed.
On opening the stopcock beneath water, the fact that the gas
has been absorbed will be seen by the water rushing into the
flask. The amount of bromine to be used must be carefully
arranged so that there is a slight excess of ethylene, in which
case the colourlessness of the ethylene bromide will be seen. If
an excess of bromine be employed, the absorption will be more
complete, but the colour of the bromine will not be removed.
From 6 to 6½ grams of bromine to the litre of ethylene will give
a satisfactory result.

462. The absorption of ethylene by fuming sulphuric acid
may be shown by introducing the gas into a tube standing over
mercury, the tube being furnished at its upper end with a
stoppered funnel. A small quantity of the fuming acid is
allowed to enter the tube and slowly to trickle down the side of
the glass, when the gas will be gradually absorbed.

ACETYLENE. C_2H_2

463. Synthetical formation, by the union of hydrogen with carbon in the electric arc. A glass globe having three tubulures is fitted up as seen in the figure. Two stout wires, to which are attached carbon rods, are passed through two corks placed opposite to each other, the wires being sufficiently easy in the corks to allow of the carbons being moved from the outside. Through one of these corks there also passes a glass tube, which serves as the exit tube for the gases ; the other neck is occupied by a cork through which a tube passes, and by means of which hydrogen gas is led into the apparatus. A small quantity of

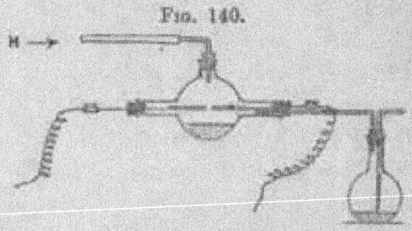

Fig. 140.

sand should be placed in the globe, to prevent any fragments of heated carbon which drop from cracking it. The gas as it leaves the globe is passed through a flask containing ammoniacal cuprous chloride, the tube delivering the gas ending just below the cork, while the exit tube reaches nearly to the surface of the liquid. A stream of hydrogen is passed through the apparatus, and when the air is expelled the two carbons are brought together and a small electric arc, from the current yielded by twenty Grove's cells, is maintained within the globe. The acetylene so produced is swept by the hydrogen into the flask containing the copper solution, and there manifests its

presence by the formation of the red cuprous acetylide.
$Cu_2Cl_2, 2NH_3 + H_2O + C_2H_2 = 2NH_4Cl + C_2H_2Cu_2O.$

464. The solution of cuprous chloride may be prepared as
follows:—A quantity of strong hydrochloric acid is boiled in a
flask, and copper oxide added in about the proportion of 10 grams
to 100 cubic centimetres of acid. A quantity of copper clippings or turnings is then introduced, and the solution boiled for
about half an hour in contact with the excess of copper. The
solution of cuprous chloride so obtained is largely diluted with
water, and the white precipitate of cuprous chloride allowed to
settle, and washed once or twice by decantation. It is then dissolved by adding to it a strong solution of ammonium chloride;
the solution so obtained will probably be slightly brown in
colour, but by acidulating with a few drops of hydrochloric acid
and placing in the solution a few strips of metallic copper, the
solution will soon become colourless, and may be preserved. For
use the solution is rendered alkaline by adding a few drops of
strong ammonia.

A stock of the cuprous chloride may be conveniently obtained by pouring the solution of copper oxide in hydrochloric
acid into a large bottle previously one-third filled with copper
clippings, and allowing the whole to stand for some days or
weeks, the bottle being closed with a caoutchouc stopper. The
solution soon becomes nearly colourless, and a thick deposit of
cuprous chloride gradually accumulates at the bottom; a quantity of this may be taken as required and treated in the manner
described.

465. Preparation from calcium carbide. $CaC_2 + 2H_2O$
$= CaH_2O_2 + C_2H_2.$ Acetylene is most readily obtained by
acting upon calcium carbide (which is now an article of
commerce) with water. A few fragments of the carbide are
placed in a dry flask which is provided with a dropping funnel
and delivery tube. Water is allowed slowly to enter, drop by
drop, when a rapid and steady evolution of acetylene at once

takes place. Large quantities of the gas can in this way be prepared in a few minutes.

466. **Preparation by the action of alcoholic potash upon ethylene bromide.** $C_2H_4Br_2 + 2KHO = 2KBr + 2H_2O + C_2H_4$. A small wide-necked flask is fitted with a cork carrying a dropping funnel and a reflux condenser; a delivery tube from

FIG. 141.

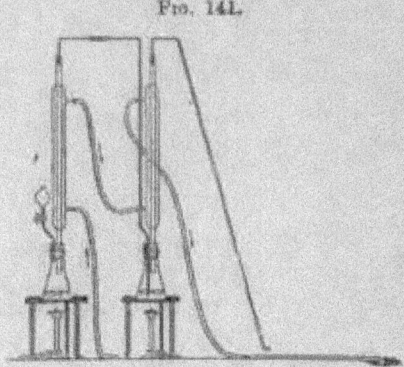

the upper end of the condenser passes into a second similar flask also carrying a reflux condenser, this tube reaching nearly to the bottom of the flask. A second delivery tube carries the gas to the pneumatic trough (see fig. 141).

FIG. 142.

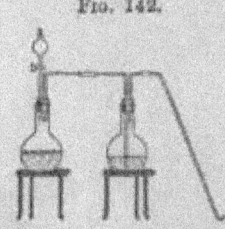

Alcoholic potash is gently heated in both flasks, and ethylene bromide allowed to drop into the first. The gas which is produced carries with it the vapours of alcohol and ethylene bromide, a portion of which condenses in the first reflux condenser and is returned to the flask. Such ethylene bromide as is carried forward is decomposed by being made to bubble through the alcoholic potash contained in the second flask, and the gas which leaves after being deprived of alcohol by the second condenser may be collected at the pneumatic trough, and will be found to be pure acetylene. If only small quantities of the gas are required the reflux condensers may be dispensed with and the simpler apparatus shown in fig. 142 may be substituted.

467. **Preparation from cuprous acetylide.** $C_2H_2Cu_2O + 2HCl = Cu_2Cl_2 + H_2O + C_2H_2$. A quantity of freshly-prepared and undried cuprous acetylide is placed in a small flask fitted with a delivery tube. The flask is then filled nearly to the cork with hydrochloric acid. On gently warming the solution effervescence sets in and acetylene is evolved.

468. **To show the formation of acetylene by passing electric sparks through marsh gas.** A wide glass tube, about 20 centimetres long, is fitted at each end with a cork through which pass a short glass tube and a stout metal wire. A gentle current of marsh gas is passed through the tube and a stream of electric sparks is discharged between the wires. The gas, as it leaves the tube, may be passed into a flask containing the ammoniacal cuprous chloride, arranged as in Experiment 463.

469. **To show the formation of acetylene when a Bunsen lamp burns at the base.** A dry cylinder is held for a few moments over the chimney of a Bunsen that is alight below. The cylinder is then covered with a glass plate and a small quantity of the solution of cuprous chloride in ammonium chloride is poured in, so as to moisten the sides of the vessel. A few drops of strong ammonia are then introduced, the ammonia being delivered with a pipette, and made to run down a part of the cylinder that is dry; sufficient ammonia gas will be evolved to render

alkaline the film of cuprous chloride solution which is moistening the sides of the cylinder, and this immediately absorbs the acetylene present and forms the red compound.

470. The presence of acetylene in ordinary coal gas may be shown by allowing the gas to bubble for a few minutes through ammoniacal cuprous chloride contained in a small wash bottle or in a U tube.

471. To show the formation of acetylene during the incomplete combustion of coal gas, which results from burning a jet of air in an atmosphere of coal gas. A jet of air is caused to burn in coal gas, and the products of the combustion aspirated through ammoniacal cuprous chloride contained in a cylinder. For this purpose the apparatus described under 'Combustion,' Experiment 361, may be used. A glass tube bent at right

Fig. 113.

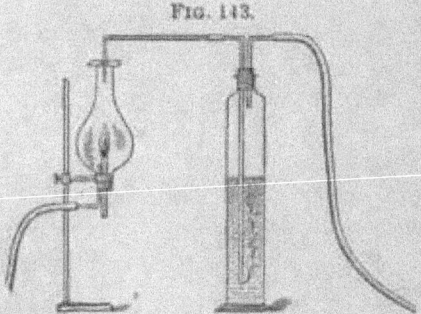

angles is introduced into the chimney through the hole in the asbestos card, and is connected at the other end to the absorbing vessel, which is in connection with any convenient aspirating apparatus, such as a Bunsen's pump. The gaseous contents of the lamp chimney are in this way made to bubble through the copper solution, where the acetylene is absorbed.

With this apparatus large quantities of the copper compound may be prepared, the copper solution being contained in a large Woulf's bottle, and the operation allowed to continue for some hours. Two or three tubes, leading to as many separate absorbing bottles, may at the same time be inserted into the chimney, so long as the united volume of gas which they draw away is not greater than the volume of coal gas which is passing through the apparatus, in which case air would be sucked down through the hole in the card, and passing through the copper solution would oxidise it. It is easy to ascertain whether there is an excess of coal gas escaping either by applying a light, or better by bringing a taper which has been lighted and then blown out, and is still smoking, near to the hole, and observing whether the smoke is carried up, or drawn into the chimney. The cuprous acetylide, so prepared after being thoroughly washed and carefully dried, may be preserved for years. It should be kept in a bottle, the stopper of which is greased, and care should be taken to prevent any of the solid from getting between the stopper and the neck, as the friction caused by removing the stopper is liable to cause the explosion of the substance.

472. For the liquefaction of acetylene see Liquefaction of Gases, Experiment 654.

FLAME, AND LUMINOSITY OF FLAME

473. To show the deposition of solid matter from luminous flames. From a flame of burning coal gas. A white plate or porcelain dish is depressed upon a ordinary gas flame, when a black deposit of soot is at once obtained.

A candle flame, or the flame of any burning hydrocarbon, may be substituted for the coal gas.

474. From a flame of burning nickel carbonyl. This gas burns with a highly luminous flame, and when a white plate is depressed upon the flame, a black deposit of nickel is obtained. For the preparation of nickel carbonyl see Experiment 446.

475. To show the deposition of solid matter from non-luminous flames. From a flame of antimoniuretted hydrogen. This gas burns with a flame which is practically non-luminous, and when a cold plate is depressed upon the flame a black deposit of antimony is obtained.

The same is the case when arseniuretted hydrogen is substituted for antimoniuretted hydrogen.

476. To show the separation of solid particles in a gas flame by the formation of vortices in the flame. When two coal-gas flames, burning from the ends of straight glass jets, are made to impinge upon each other in the manner shown in fig. 144, the

FIG. 144.

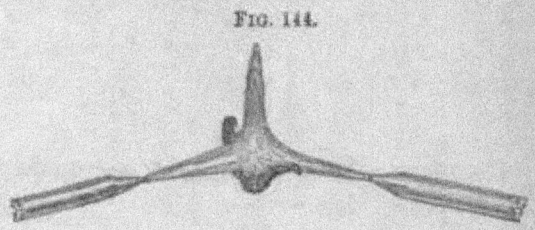

under part of the flame will be seen to curl over and extend on each side in the form of two horns. By carefully adjusting the size of the flames, and the angle at which they impinge, these portions of the flame may be made to take up a rotating motion, when solid particles will be seen to form in the flame, and will spin round as in a vortex, burning with little sparkles as they are thrown out of the flame.

477. A more striking method of illustrating the same phenomenon, and one more easily under control, is seen in fig. 145.

An ordinary Argand burner is provided with a chimney 9 centimetres high. The chimney is covered with a piece of fine wire gauze (15 meshes to a centimetre), and an ordinary mouth blowpipe, supported in a clamp, is so arranged that the fine end presses gently against the wire gauze, in a vertical direction.

The Argand is lighted and the flame turned down, until only the faintest blue ring of flame is left, the object of this flame being merely to create a steady gentle draught up the chimney. Coal-gas is then delivered through the blowpipe jet. The gas will be ignited by the Argand, and when regulated the

Fig. 145.

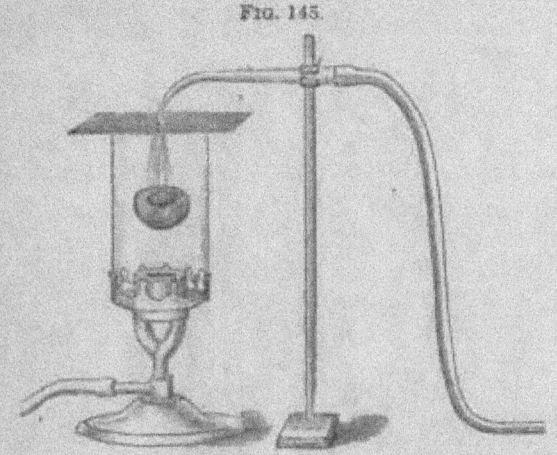

flame assumes the shape of a bowl, or vortex ring. Solid particles will be seen revolving in the vortices, and, owing to the rather feeble luminosity of the flame, the sparkles caused by their combustion produce a very beautiful effect.

478. **Luminous flames in which no solid matter is present.** A number of highly luminous flames can be obtained in which, from the nature of the products of combustion, and their known volatility, no solid matter can be present.

The flame of hydrogen burning in oxygen under pressure. This is most readily shown by exploding a mixture of oxygen and hydrogen in a 'Cavendish' endiometer. (See Water, Experiment 73.) The combustion of the two gases is attended with a dazzling flash of light.

479. The flame of phosphoretted hydrogen, burning in air or in oxygen. (See Experiment 547.)

480. The flame of carbon disulphide burning in oxygen. (See Experiment 614.)

481. The flame of a burning mixture of carbon disulphide and nitric oxide. (See Experiment 319.)

482. The flame of burning arsenic in oxygen. (See Experiment 622.)

483. The effect of pressure in influencing the luminosity of flames may be illustrated by burning a candle under diminished pressure. For this purpose a burning candle is placed beneath a large air-pump receiver, and the pressure quickly reduced by a few strokes of the pump. To obviate as far as possible the effect of the carbon dioxide produced by the burning candle, a shallow dish, as large as possible, is filled with slaked lime and previously introduced beneath the bell-jar. As the pressure is diminished, the luminosity of the candle will be seen to rapidly decrease; but, on allowing air to re-enter, the flame will once more assume its original condition. Care must be taken not to reduce the pressure too far, or the flame will be extinguished. It is well to connect with the pump a long tube dipping into mercury, that the exact pressure in millimetres may be indicated, or to place a small aneroid barometer beneath the air-pump receiver.

484. By burning arsenic under diminished pressure. A large two-necked Woulf's bottle, capable of holding six or eight litres, is connected with an air-pump by means of a piece of wide glass tube about 30 centimetres long, which is loosely packed with cotton wool. A similar glass tube, about 40 centimetres long, and in which the arsenic is to be burnt, is fitted to the second neck of the Woulf's bottle; this tube is fitted with a caoutchouc cork carrying a piece of fine lead pipe which is connected with an oxygen gas bottle. At a convenient part of the

apparatus a gauge must be introduced, which may consist of a glass tube, not less than 85 centimetres long, dipping into a small mercury trough, and having a rough cardboard scale attached. The apparatus is exhausted as far as possible, and the arsenic gently heated at one point by a Bunsen flame; oxygen is then allowed slowly to enter, when the arsenic will ignite, but will be seen to burn with a somewhat feebly luminous flame.

As the arsenic burns, the pressure may be kept down by further pumping. The products of the combustion will for the

Fig. 146.

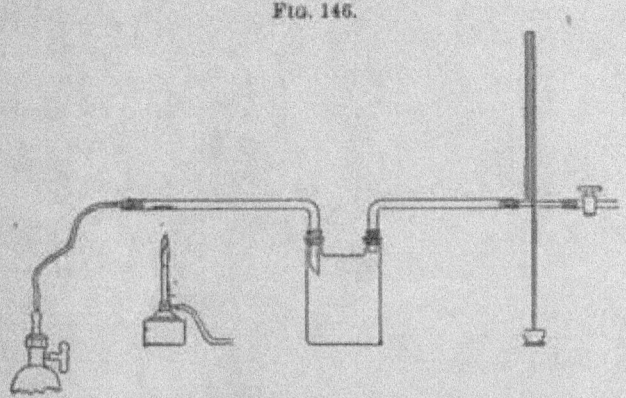

most part condense in the Woulf's bottle, and the cotton wool filter will prevent any arsenious oxide from entering the pump.

The luminosity of the flame of arsenic burning in oxygen at the ordinary pressure may be compared. (See Exp. 622.)

485. By burning an alcohol flame under increased pressure. The flame of burning alcohol may be rendered distinctly luminous by increasing the pressure of the air in which it is burning. For this purpose a large and strong glass globe with two necks, as shown in the figure, is fitted with two caoutchouc corks which can be securely wired into the necks. Through the lower cork a spirit lamp is passed, made of a short piece of wide

glass tube closed at one end, and fitted at the other with a cork
carrying a short glass tube through which the wick passes; the
cork must either be pierced with a second hole or have a small
slot cut longitudinally upon it, to allow of free communication
between the outside atmosphere and the interior of the little
lamp.

Through the same cork which carries the lamp is passed a
piece of narrow composition or lead pipe; this pipe should pro-

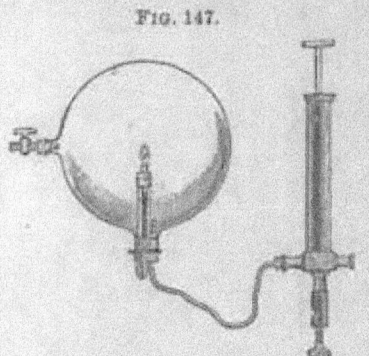

FIG. 147.

ject 3 or 4 centimetres through
the cork; the end should be
closed by being gently hammered up, and two or three
small holes pierced in the side;
the object being to prevent the
flame of the lamp from being
disturbed by the draught which
would be caused if air were
driven into the vessel through
the open end of the pipe. This
lead pipe is connected to a
small condensing pump. The second neck of the globe is
provided with a stopcock.

The lamp may be supplied with methylated spirit to which
a small quantity of amyl alcohol (fusel oil) has been added, in
order to produce a flame which will show a small luminous tip.

The lamp may be lighted by introducing a taper through
the side neck of the flask; the cork is then quickly secured, and
the air rapidly compressed by working the pump. The flame
will be seen to grow considerably more luminous as the pressure
is increased, and on allowing the air to escape by the stopcock
the flame will once more assume its original condition.

486. The combustion of a mixture of oxygen and hydrogen
in the Cavendish eudiometer, where at the moment of the explosion the pressure is enormously increased, may be contrasted
with the combustion of the same mixture when unconfined, as

in soap bubbles. A mixture of the two gases contained in a small gas-holder is made to bubble through a solution of soap contained in a mortar, until the vessel is filled with a froth of bubbles. The delivery tube is then removed, and in a darkened room the mixture exploded, when it will be seen that there is no luminous flash attending the combustion.

The froth should be ignited by means of a small spirit flame, which may conveniently be obtained by plugging a fragment of cotton wool into one end of a piece of glass tube, which is then dipped into alcohol and ignited.

487. The effect of increased temperature upon the luminosity of flames may be illustrated by heating coal-gas to redness before its combustion. This may be shown by elongating the tube of a Bunsen lamp by means of a platinum tube about 12 centimetres long, and of such a bore that it will just fit upon the Bunsen. The lamp is supported by a clamp in a position shown in the figure, and the

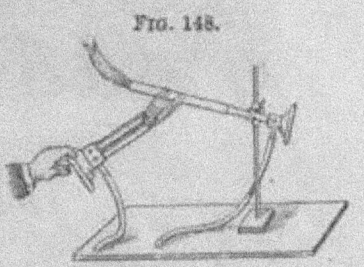

Fig. 148.

supply of air so adjusted that the lamp burns with a flame which just begins to show luminosity at the tip. On heating the platinum tube by means of a powerful Bunsen, the flame will be seen to increase greatly in luminosity.

488. By heating both the coal-gas and air to redness before combustion. This may be illustrated by means of the apparatus already described (Marsh Gas, Experiment 453). A flame of coal-gas, burning under the ordinary conditions, and of the same size as that obtained in the furnace, should be placed alongside for comparison.

489. A ready method for roughly comparing the luminosity of different flames, suitable for lecture purposes, is the shadow test. For this purpose a rod, conveniently a retort stand, is

placed near to a small transparent screen (either made of tracing-paper stretched over a light wooden frame, or, better, a sheet of opal glass), and the two lights which are to be compared are so arranged that the shadows of the rod cast by each flame fall

FIG. 149.

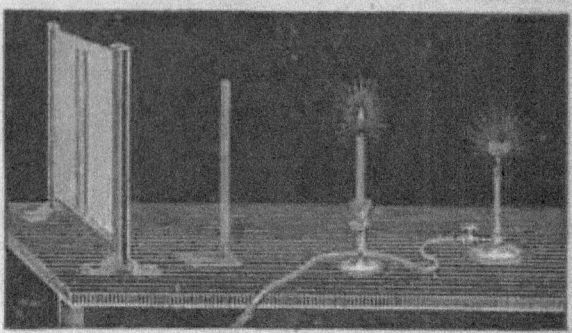

in close juxta-position upon the screen (fig. 149). The feebler light is then moved nearer to the screen until the two shadows are of equal intensity. The relative luminosity of the two flames will be inversely as the square of their distances from the screen, which is ascertained by a measure.

490. The increased luminosity obtained by heating marsh gas and air to redness may also be shown. (See Exp. 453.)

491. By burning boiling phosphorus in red-hot chlorine. When phosphorus is introduced into chlorine gas, it spontaneously ignites and burns with a flame of very feeble luminosity; but if the two elements are strongly heated before combination, the luminosity is greatly increased. To effect this, a bulb is blown near to one end of a combustion tube, which is heated to redness in a furnace. Upon the end of the tube is fitted a cork, over which a wide glass tube about 30 centimetres long will easily pass, the open end of which delivers into a draught flue. The other end of the combustion tube is connected to a chlorine apparatus, and a stream of gas passed through the apparatus.

When all the air has been displaced, the passage of chlorine through the tube is stopped; the wide tube is then withdrawn and a piece of phosphorus pushed into the bulb by means of a glass rod, and the tube quickly replaced. The phosphorus is then heated by means of a Bunsen flame, and when it is in active ebullition the stream of chlorine is again diverted into the tube; the red-hot chlorine thus coming in

FIG. 150.

contact with the boiling phosphorus, the latter element will be seen to burn with a comparatively high degree of luminosity.

It is well to cover the wide tube with black paper, so that, if any air is drawn in between the cork and the tube, the light produced by its combination with the phosphorus vapour shall not be seen, and so mask the effect of the combustion in the bulb.

492. To show the diminution in luminosity of a coal-gas flame by cooling it. A cold block of iron, conveniently an ordinary domestic 'flat-iron,' is brought against a flame of gas burning from a fish-tail burner in such a way that the flame just plays against the cold metal; it will be seen that the luminosity is almost entirely destroyed.

FIG. 151.

493. This experiment may be modified in the following

Q

way:—The flat-iron is replaced by a stout platinum dish, which is supported by a clamp in the position shown in fig. 152. A

Fig. 152.

small gas flame, burning from a fine jet, is made to impinge against the dish, whereby the luminosity of the flame is nearly destroyed. A small blowpipe flame is then directed into the dish so as to strongly heat the metal at the point where the outer flame impinges upon it. As the platinum becomes hot the luminosity of the flame is restored very nearly to its original brightness.

494. The shadows cast by flames. When certain luminous flames are placed between a strong light and a screen, it will be seen that they cast a distinct shadow, which will be found to be coincident with the luminous portion of the flame; this fact

Fig. 153.

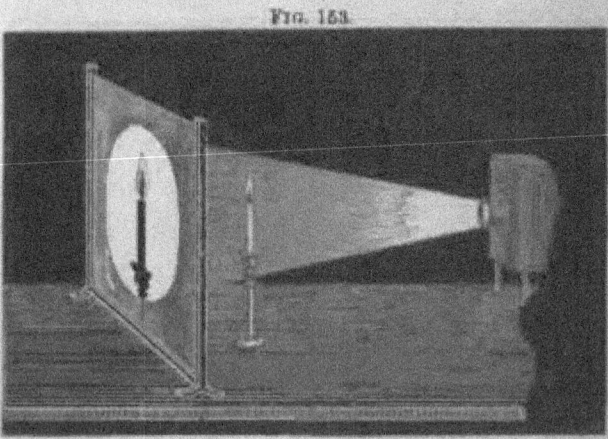

is considered by some chemists as a strong argument in support of the solid particle theory. The phenomenon may be shown with either a candle flame or a flame of burning coal-gas. A strong ray from a naked light is allowed to fall upon a small transparent screen either of tracing-paper or opal glass, and a lighted candle placed in the path of the beam so that its shadow may fall upon the screen (fig. 153); there will be seen in the central part of the shadow of the flame a dark shadow; by holding a pair of dividers against the luminous portion of the candle flame, it will be seen that this dark shadow is coincident with it. If the flame be disturbed by slight draughts and so made to smoke, the shadow of the luminous portion will be seen to extend into the shadow cast by the smoke. A flame of coal-gas burning from a straight jet shows the same phenomenon as a candle flame. If the gas be burnt from a fish-tail burner, and the flame be placed broadside to the screen, no shadow is observed; on turning the flame edgeways to the screen it is of sufficient thickness to cast a very distinct shadow.

495. Instead of casting a shadow by means of a naked light, which shadow must of necessity be rather small and therefore

FIG. 154.

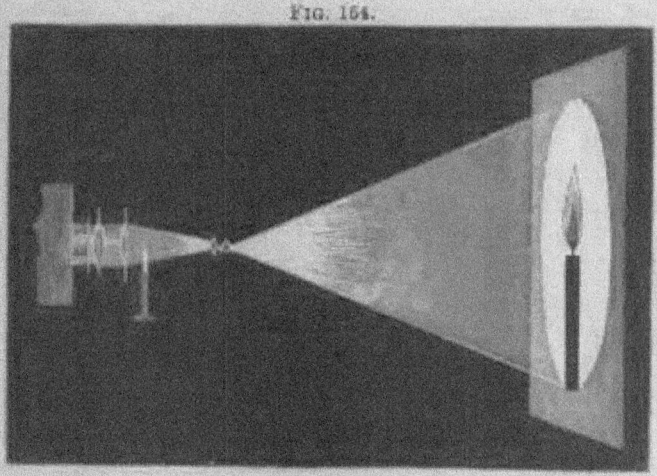

not easily seen by a large audience, it is possible to project a magnified image of the flame upon the screen in the usual manner by the use of lenses.

The combination of lenses most suitable for the purpose is described under Lantern Illustrations, and if a right-angle prism be employed as shown in the figure, the image of the candle and flame will be reinverted. Besides the fact that in this way the phenomenon may be rendered more visible to an audience, this arrangement has the additional advantage of giving no image of the flame, and heated current of gases rising from it, resulting from differences in refraction. In the case of the simple shadow above described, the shadow of the whole flame is almost as pronounced as that of the central luminous part, which appears as a darker portion in the midst of the flame. By the use of the lenses this effect, due to refraction, is so much diminished as to be hardly visible.

496. A large number of luminous flames can be produced which do not cast any shadow when placed in a strong light as above described.

(1) The flame of carbon disulphide burning in oxygen (see Carbon Disulphide, No. 614).

(2) The flame of phosphoretted hydrogen burning in air, or with still greater luminosity in oxygen.

(3) The luminous flame of hydrogen charged with chloro-chromic acid. This is produced by passing hydrogen through a small flask containing a few cubic centimetres of chloro-chromic acid, obtained by distilling a mixture of sodium chloride, potassium bichromate, and sulphuric acid.

(4) The flame of coal-gas itself when burnt in oxygen. A similar apparatus to that employed for Nos. 1 and 2 may be used, and it will be seen that as the flame is surrounded with oxygen, although its luminosity increases, its shadow gradually vanishes.

(5) The flame of hydrogen or carbon monoxide impregnated with nickel carbonyl (see Carbon Monoxide, No. 446).

497. The Bunsen flame. The two chief principles involved in the construction of the Bunsen lamp are briefly the following. First, that when a stream of gas is made to issue from a small

orifice a partial vacuum is produced in the immediate neighbourhood of the jet; and, second, the relation between the velocity at which the gases issue from the lamp and the rate of propagation of combustion in the mixture of air and coal-gas.

498. To illustrate the fact that the escape of a gas through a narrow jet causes a reduction in pressure, a number of experiments may be arranged. A steel gas-bottle containing either oxygen or hydrogen under great pressure is furnished with  FIG. 155. a jet made by soldering upon the ordinary nozzle a stout brass plate, about 5 centimetres diameter, in the centre of which a small hole has been drilled. The bottle is supported in a vertical position with the jet directed downwards. The valve of the bottle is opened wide so that the gas may rush from the jet with its full force, and a disc of metal, either iron or brass, about 3 centimetres diameter and 1 centimetre thick, is held in the jet, when it will be seen that, instead of being blown away by the rush of gas, it is held up by the partial vacuum which exists in the area surrounding the jet.

499. The same phenomenon may be shown by a slight modification of the experiment whereby a very much less pressure of gas is required. A circular glass plate, 10 centimetres in diameter,

having a hole in the centre, is fitted upon a cork through which passes a straight piece of glass tube, the end of which has been nearly closed up by being heated in a blowpipe flame until the orifice is about 2 millimetres in diameter.

FIG. 156.

The end of the tube and the cork are made flush with the surface of the glass plate. Three small pieces of cork, about 8 millimetres thick, are cemented upon the under surface of the plate as feet, in order that the apparatus may be stood over a disc of cardboard without touching it. If while it is standing in position a stream of gas be driven through the tube (either by blowing with the mouth or by means of compressed gas from a bottle) the cardboard disc will be lifted from the table and will adhere to the glass. The instant the current of gas is stopped the disc falls. It is much better to employ a stream of gas from a bottle of compressed hydrogen or oxygen than to blow with the mouth, as in the latter case the moisture accompanying the breath is apt to curl the card, besides wetting the glass.

500. The same phenomenon may also be illustrated by causing a strong jet of gas to escape immediately above a long glass tube which dips into water, when this area of reduced pressure will be manifested by the rise of water in the tube. For this purpose a long glass tube drawn to a jet at one end and bent at right angles at a point about 15 centimetres from the end is cemented into the nozzle of a bottle of compressed gas, either oxygen or hydrogen (fig. 157). A second tube also drawn to a blunt point is attached as shown in the figure, so that its end is near to and slightly below the point of the other tube. This tube is made to dip into water. Upon opening the valve and causing a stream of gas to blow across the orifice of the vertical tube, the water will be drawn up to the point, and forcibly scattered as a fine spray.

FIG. 157.

501. A convenient apparatus for showing smaller effects of this order may be constructed in the following way. A bulb, blown upon a piece of glass tube about 12 millimetres bore, has a branch tube bent as shown in the figure. To this, by means of a small piece of caoutchouc tube, a long piece of glass tube is attached. Into the bulb-tube

is inserted a glass jet reaching rather more than half-way into the bulb. The apparatus is filled with coloured water so that the jet projects above the liquid, and the long tube is supported so that the water reaches nearly to the open end. Upon blow-

FIG. 158

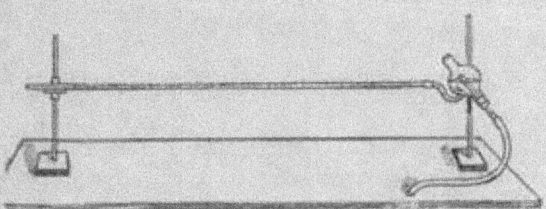

ing with the mouth through the jet, by means of the flexible tube, the same area of reduced pressure will be created round the jet, which will cause the coloured liquid in the tube to be drawn along towards the bulb.

502. By means of a small manometer of this description it may be shown that in the Bunsen lamp this area of reduced pressure exists in the neighbourhood of the small jet at the base of the chimney, and that it is owing to this fact that air is drawn in through the 'air-holes.' A brass tube about 3 centi-metres long is soldered into one of the air-holes of a Bunsen lamp, and the remaining hole or holes stopped up. Into the brass tube is fitted, by means of a cork, the end of a small mano-meter somewhat similar in con-struction to that above described, and containing coloured water (fig. 159). The apparatus is placed upon a levelling block (see fig. 45), and supported in front of the lantern in such a position that an image of the small straight glass tube can be projected upon the screen, the liquid being

FIG. 159.

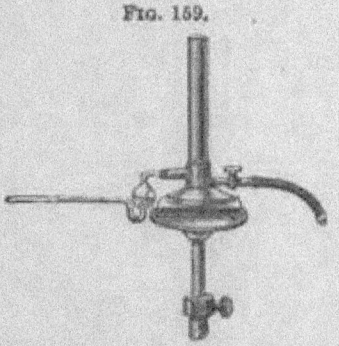

brought into the field by means of the levelling screws. The Bunsen is connected with the ordinary gas supply, and, as soon as the gas is turned on, a movement of the liquid in the tube will at once be seen to take place in the direction towards the bulb; on turning the gas off, the liquid will return to its former position.

The turning on and off of the gas should be done at the supply and not by the cock upon the lamp, in order to avoid disturbing the apparatus.

503. To show that in order that a Bunsen lamp shall burn there must be a certain relation between the velocity of the issuing gases and the rate of propagation of combustion in the mixture of air and gas. A lighted Bunsen may be slowly turned down; when the velocity of the issuing gas has been retarded to a certain point, the flame will be seen to 'strike down' the chimney; that is to say, the combustion of the mixture of air and gas will be propagated through that volume of it which

Fig. 160.

is at the moment in the chimney, and the gas will be ignited at the small jet within the lamp.

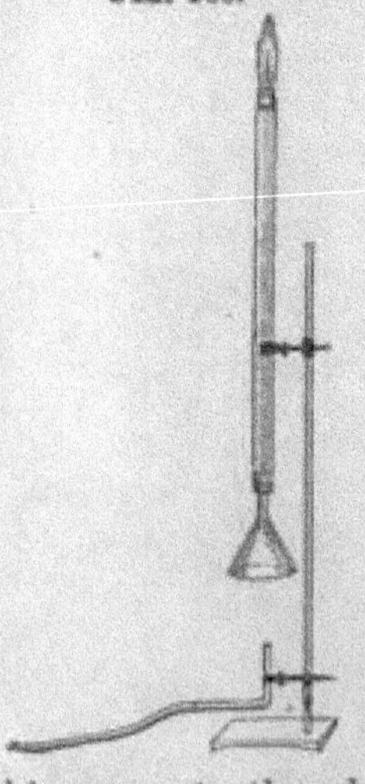

504. This experiment may be shown in a more striking manner by means of the following device. A glass tube about 60 centimetres long and 2 centimetres bore is fitted on to a funnel as shown in the figure. Coal-gas is delivered through a small glass tube supported beneath the funnel, and the mixture of gas and air is inflamed as it issues from the top of the tube. A piece of platinum foil should be inserted in the top of the glass tube, so as to form a short platinum tube projecting about a centimetre above the glass; this prevents the glass from being cracked by the heat of the flame. If now the velocity of the issuing gases be checked by gradually reducing the supply of coal-gas, a point will be reached when the flame will recede down the tube. By a little adjustment the flame may be caused to travel down either rapidly, or so slowly that its progress can be easily watched.

505. If the long glass tube be slightly choked at any point, the upward velocity of the gases at the constricted part will be increased, and the receding flame may be made to stop in its descent down the tube at this point where the upward velocity of the gases is greater. This may be shown by introducing into a tube similar to fig. 160 a roll of asbestos cardboard, 2 millimetres thick, forming a short tube 3 or 4 centimetres long. This is pushed down about 12 centimetres from the top. A mixture of air and gas is passed up the tube as described above, and when the gas is so adjusted that the flame slowly descends, it will be seen that it stops in its downward movement when it reaches the asbestos, and there remains stationary (fig. 161). It will be seen that the entire flame does not recede down the tube, but only that part which appears as

Fig. 161.

the inner cone; the outer cone or skin of the flame remains still burning at the top, although rendered very languid on account of the products of the combustion of the other.

506. This experiment may be rendered more under control, and the phenomenon better studied, by a slight modification of the apparatus, whereby the supply of air can be regulated as well as that of the gas.

For this purpose a glass tube, 50 centimetres long and 15 millimetres bore, is fitted at one end with a cork carrying a T piece, preferably of metal. This tube is surrounded by a second tube 22 millimetres bore, which is made to slide easily up and down over the inner tube by means of two rings of asbestos cardboard (c c, fig. 162) which are cemented upon the inner tube. Both tubes are provided with platinum ends as described above.

The outer tube is lowered until the top of it is level with that of the inner one, and coal-gas is passed in by one limb of the T tube and inflamed as it issues; a stream of air is then driven in through the other limb, the flame at once is converted into the ordinary non-luminous Bunsen flame; by regulating the supply of air and coal-gas the flame can readily be brought into a condition in which it will nearly recede down the tube.

FIG. 162.   FIG. 163.

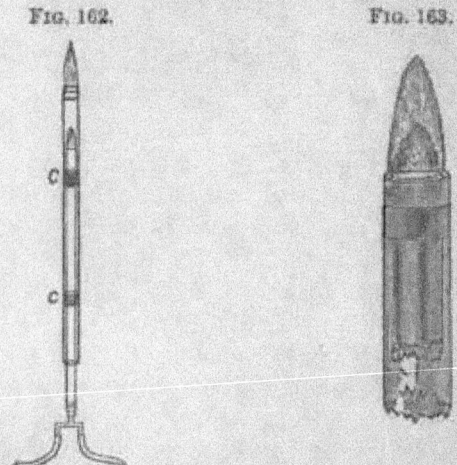

It then presents the appearance shown in fig. 163, the inner cone being greatly reduced in height, and appearing of a curious greenish colour. If, while the flame is in this condition, the outer tube be raised, the outer mantle of the flame will be lifted away from the inner cone, and the latter will continue to burn at the orifice of the inner tube, while the outer flame is burning above it (fig. 162). If, now, the proportion of air to gas be disturbed, the inner flame can be made either to ascend to the mouth of the upper tube, and the ordinary Bunsen flame reproduced, or to recede down the inner tube to the bottom; by increasing the coal-gas the inner flame ascends, and by increasing the air supply it descends.

The air for this experiment is best supplied from a large gasometer, capable of holding several cubic feet,[1] but in the absence of this a sufficiently regular stream may be obtained by a combination of an ordinary foot-bellows, as used for a blowpipe, with a small gas-holder. The arrangement is shown in fig. 164. The tube A is in connection with the bellows, and delivers air into the gas-holder, which contains water to a depth of about 10 centimetres. The air leaves the holder by the second cock, and is passed through a T tube, one limb of which dips into mercury, the other being connected to the apparatus where the flame experiment is to be performed. The depth of mercury in the bottle must be such that, before air bubbles through, the water in the gas-holder has been forced up into the top reservoir to a depth of about 7 centimetres. A plug of cotton wool thrust

FIG. 164.

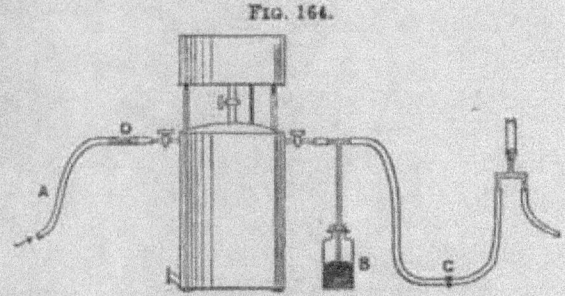

into the neck of the mercury bottle will prevent any splashing of the metal over the lip. The stream of air delivered into the flame apparatus is regulated entirely by the screw-cock C.

Between the bellows and the gas-holder a valve should be introduced (D, fig. 164) in order to relieve the bellows from the pressure, and prevent leakage back in that direction. Such a valve may be constructed as shown in fig. 165. F is a short piece of brass tube upon the end of which is soldered a disc with

[1] 1 cubic foot = 28·3 litres.

a hole in it rather less than the diameter of the tube. Over this is tied a strip of oil silk, and the whole fixed inside a glass tube G by means of a cork. This will allow air to pass only in the direction indicated by the arrow.

A number of flames may be observed by means of the apparatus above described, *e.g.* mixtures of coal-gas with oxygen

FIG. 165.

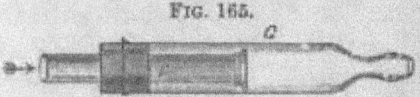

or coal-gas with nitrous oxide will produce flames which behave very similarly to the one above treated of, and, these two gases being both readily obtained compressed in cylinders, a regular stream of them can be produced without difficulty.

In the same way ammonia gas mixed with either oxygen or nitrous oxide will burn with a flame which may be separated in the same way into two parts.

507. To show that a Bunsen flame is hollow, by introducing a match into the inner cone of the flame. A pin is pushed through a match, at about 3 centimetres from the tip, and the match suspended in the chimney of an unlighted lamp, in the manner shown in the figure. The gas is then lighted, when it will be seen that the match will remain suspended within the flame without becoming ignited. If the flame be disturbed by a draught, or by gently fanning it, so as to cause the walls of the flame to touch the tip of the match, inflammation of the latter at once takes place.

FIG. 166.

508. This fact may also be illustrated by depressing upon the flame a sheet of white paper, when a charred ring will be produced upon the paper.

For this purpose a well glazed and not too thick foolscap paper should be selected, and for convenience the sheets may be cut into four parts. A Bunsen flame about

7 centimetres high, having a good air supply, should be placed where it is not exposed to much draught, and a piece of the paper quickly depressed well down upon the flame. The paper rapidly begins to char, and the instant the ring is distinctly visible the paper must be quickly removed. By depressing the paper upon the flame at various angles, charred elliptical sections of the flame may be obtained.

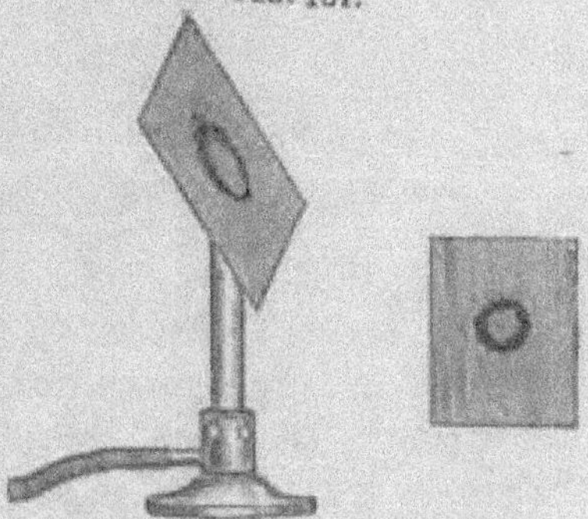

It is sometimes suggested to soak the paper in a solution of alum, and dry it, in order to render it less inflammable, but as this process is apt to wrinkle the paper, and always spoils its surface, it is not to be recommended.

FIG. 167.

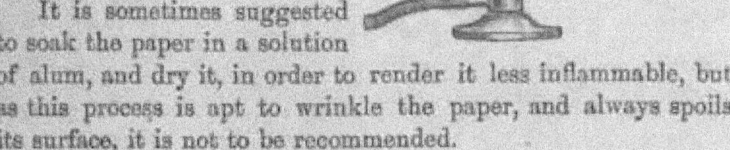

509. Instead of charring paper, a very striking method of showing the hollow character of a Bunsen flame consists in employing paper which is coated over with iodide of mercury. Upon depressing a piece of such paper upon the flame, the red iodide is instantaneously changed to the yellow variety at the part where the flame touches it, and a yellow ring is produced. The dry powdered iodide of mercury is spread upon one side of the paper, by being rubbed over the surface with the finger. As only a moderate heat is required to effect the change in the iodide, the paper must be quickly depressed upon the flame, the coated side being uppermost, and instantly withdrawn again.

FIG. 168.

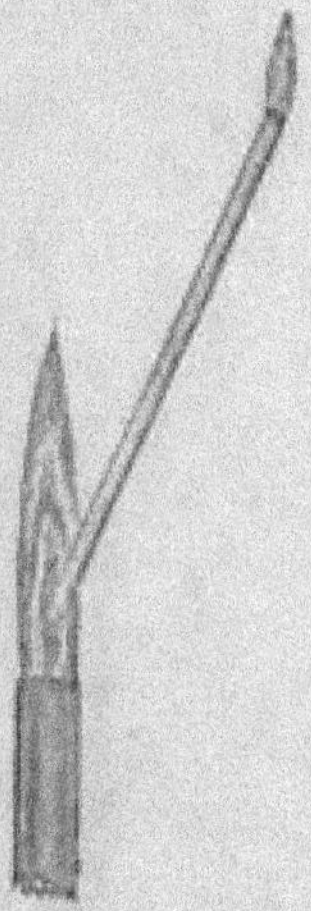

510. In order to show that the interior of a Bunsen flame consists of unburnt gases, a glass tube may be held in the flame in the manner shown in the figure. The gases within the flame will ascend in the tube, and may be inflamed at the upper end.

SILICON

511. The reduction of a silicate, by heating with metallic sodium, may be shown by heating a fragment of the metal in an ordinary test tube. In a few moments the glass becomes blackened, owing to the reduction of the silica it contains.

512. Amorphous silicon may be prepared by heating to redness in a small porcelain crucible a few grams of potassium silicofluoride, and throwing in fragments of sodium, and re-placing the lid of the crucible as quickly as possible. When cold, the mass appears black. On treating the contents of the crucible with water, and filtering the liquid, the silicon is obtained as a dark brown, almost black, powder. $K_2SiF_6 + 2Na_2 = 2KF + 4NaF + Si$.

SILICON HYDRIDE. SiH_4

513. Preparation by the action of hydrochloric acid upon magnesium silicide. $Mg_2Si + 4HCl = 2MgCl_2 + SiH_4$. A small quantity of coarsely powdered magnesium silicide is placed in a small beaker and drenched by the addition of strong hydrochloric acid. An instant evolution of silicon hydride takes place, and, the gas so evolved possessing the property of spontaneous in-flammability, each minute bubble, as it bursts, ignites and burns with a bright flash.

The magnesium silicide may be prepared in the following way:

80 grams of fused and powdered magnesium chloride are

introduced into a dry wide-mouth stoppered bottle. To this is
added:

70 grams of potassium silicofluoride,

10 grams potassium chloride,

10 grams of sodium chloride, and 40 grams of sodium cut
into small pieces, and the whole shaken together.

A clean Hessian crucible is heated to bright redness in a
furnace, and the mixture thrown into it. The lid is replaced as
quickly as possible, and the whole allowed to go cold. The
crucible may then be broken up, and the slag containing
the silicide coarsely powdered and preserved in a bottle the
stopper of which is greased with vaseline.

A ready method for preparing a small quantity of impure
magnesium silicide consists in heating sand with magnesium
filings. A small quantity of white sand, which has been
previously rendered perfectly dry by being heated on an iron
tray, is mixed with four times its weight of magnesium filings,
and the mixture heated strongly in a piece of hard glass tube
closed at one end. When a particular temperature is reached
the entire mass glows brightly. When quite cold the tube is
broken up, and on adding hydrochloric acid to the residue there
is generally sufficient silicon hydride produced to spontaneously
ignite, and to inflame the hydrogen which is simultaneously
evolved.

SILICIC ACID. H_4SiO_4

514. If hydrochloric acid be added to a solution of sodium
silicate, silicic acid in the gelatinous condition is precipitated.
If, on the other hand, the silicate be added to the acid, the latter
being always in considerable excess, no precipitation takes
place, the silicic acid remaining in solution. Strong hydrochloric acid should be used, and the strength of the solution
of the silicate so adjusted that when portions of the same

solution are employed, in the one case the precipitated silicic acid is obtained, and in the other the solutions remain clear.

$$Na_4SiO_4 + 4HCl = 4NaCl + H_4SiO_4.$$

If the clear solution be subjected to dialysis, a dilute solution of silicic acid free from sodium chloride and hydrochloric acid can be obtained.

515. As a lecture experiment to illustrate the process of dialysis, an emulsion of starch to which potassium iodide has been added is a convenient mixture to employ.

The solution is thrown upon a small dialyser, which is placed upon a glass tripod standing in a shallow dish of water. In a short time sufficient potassium iodide will have passed through the membrane into the water below to give the test for iodine. By adding a few drops of chlorine water, it may be shown that the starch and iodide have not both made their way through, as there will be no blueing of the mixture. On the addition of a few drops of fresh starch the blue compound will be formed, proving that the iodide alone had diffused through the dialyser.

The dialyser is readily made by means of parchment paper, which is first wetted in tepid water and then stretched over a shallow wooden hoop by means of a second similar hoop which is just large enough to slip over the first. An ordinary hair sieve will furnish the most convenient hoops for the purpose.

BORON

516. Boron in the amorphous condition may be prepared by fusing boron trioxide in a small porcelain crucible by means of a blowpipe flame, and throwing into the crucible, the lid being partially removed, fragments of sodium. Each piece of sodium deflagrates upon the surface of the fused oxide, and a black mass is formed by the reduction. The amorphous boron may be separated by treating the cooled contents of the crucible to prolonged treatment with water, and filtering the liquid.

BORIC ACID. H_3BO_3

517. Preparation by the action of hydrochloric acid upon borax. $Na_2H_2B_4O_8 + 4H_2O + 2HCl = 2NaCl + 4H_3BO_3$. To a hot and nearly saturated solution of borax a quantity of strong hydrochloric acid is added, which causes an immediate precipitation of boric acid in the form of crystalline scales. On allowing the liquid to cool, a further quantity of the acid is deposited.

518. To show the action of boric acid upon turmeric. One or two drops of strong sulphuric acid are added to a small pinch of powdered borax upon a watch glass. The mixture is then diluted with water, and the solution placed upon a piece of turmeric paper by means of a glass rod. A dark-brown stain is produced, resembling somewhat the colour produced upon turmeric by an alkali. If a drop of a solution of sodium hydroxide be brought upon the brown stain it is at once blackened.

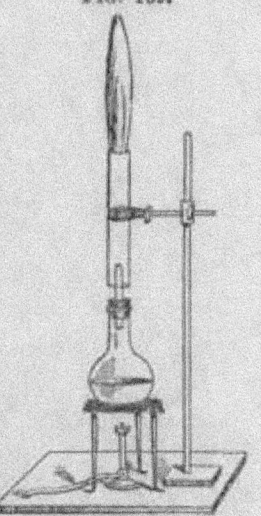

Fig. 169.

519. To show the colour imparted to a flame of alcohol by boric acid. This is most strikingly shown by placing in a small flask a few grams of finely powdered borax, and moistening it with concentrated sulphuric acid ; a quantity of methylated spirit is then added, and the flask closed with a cork carrying a short tube slightly drawn out at the end. The flask is placed upon a tripod, and a piece of glass tube 9 centimetres long and 12 millimetres bore is supported over the jet as shown in the figure.

Upon boiling the alcohol, the vapour as it issues from the small

R

jet becomes mixed with air as it ascends in the wider tube, and when inflamed at the top burns with a non-luminous flame resembling a Bunsen flame, but tinged of a brilliant green colour owing to the boric acid which vaporises with the alcohol.

PHOSPHORUS

520. Pure crystallised phosphorus may be prepared by sublimation in a vacuous tube. A small quantity of phosphorus is introduced into a tube closed at one end; the open end is then drawn out and connected to a Sprengel pump, and the tube exhausted. It is then sealed up and the phosphorus gently melted, and again allowed to solidify in the end of the tube. The object of melting the phosphorus in this manner is to prevent it from rolling about in the tube, and damaging the crystals which are produced. The upper half of the tube is then wrapped round with rag, which is kept constantly moist, and the whole placed in the dark, but freely exposed to the air, so that the evaporation of the water from the rag may continually keep the covered part of the tube slightly cooler than the rest. In this way colourless brilliant crystals of phosphorus will gradually sublime, and if the tube be kept from the light, they will retain their brilliancy and transparency unimpaired for any length of time.

521. The ready inflammability of phosphorus may be shown by rubbing a small piece between two pieces of wood which have been covered with fine sandpaper.

522. If the point of a penknife be drawn quickly two or three times over the table with a slight pressure, the friction will have raised its temperature sufficiently high that, if it be brought in contact with a fragment of phosphorus, the latter will be ignited

523. If a bar of lead or a piece of lead pipe be struck a

few sharp blows with a hammer upon an anvil or block of iron, its temperature will be raised sufficiently high to ignite phosphorus.

524. If a bar of tin, about the thickness of a lead pencil, be bent backwards and forwards briskly three or four times, sufficient heat will be developed at the point where the metal is bent to inflame phosphorus.

525. The spontaneous inflammability of finely divided phosphorus may be shown by dissolving a small quantity of phosphorus in carbon disulphide, and pouring the solution upon a filter paper, supported in a ring upon a retort stand. In the course of a few moments the disulphide evaporates and leaves the phosphorus in so finely divided a condition that it spontaneously bursts into flame.

526. This experiment may be varied by applying the solution upon a sheet of blotting-paper with a camel's hair brush, in a design, or writing a word; the time which elapses before inflammation takes place being long enough to allow of this being done. As the phosphorus inflames, the design or word is charred upon the paper.

527. To show the vaporisation of phosphorus in a current of steam. A piece of phosphorus and a small quantity of water are placed in a flask fitted with a short piece of straight glass tube, about 10 millimetres diameter. On boiling the water, the steam which issues from the jet carries with it such a quantity of phosphorus vapour that it produces a feebly luminous flame, which is best seen in a darkened room. The temperature of this flame is so low that it is incapable of igniting a match. A piece of gun-cotton may be held within it without being inflamed.

528. To show the suspended solidification of phosphorus. Two or three glass tubes, about 15 centimetres long and 5 millimetres bore, are sealed at one end, and nearly filled with clean

phosphorus. This is best accomplished by first melting beneath warm water in a small beaker a stick of phosphorus, as clean as possible. The glass tubes are filled with warm water and stood in a beaker, also containing hot water. A pipette with a long fine drawn-out end is dipped down to the bottom of the melted phosphorus, care being taken to first draw a little water into the pipette, and the clear and almost colourless liquid drawn up. The pipette is then withdrawn and the fine point introduced into the tubes to be filled, and the phosphorus allowed to run in to a depth of about 10 centimetres; about 2 centimetres of water should be left above the phosphorus. Two or three of these tubes should be prepared, and placed in a vessel of water a few degrees above the melting-point of the phosphorus.

One of the tubes is placed in front of the lamp, and an image of it projected upon the screen. After the lapse of a few minutes, during which the phosphorus will have cooled considerably below its melting-point, it is caused suddenly to solidify by touching the surface of the liquid phosphorus with a capillary glass tube, in the end of which there is a fragment of solid phosphorus. Instantly that this comes in contact with the liquid phosphorus, the mass solidifies throughout the entire tube, and the change is evident upon the screen by the transparency of the liquid being destroyed. After a very short time, however, the mass again becomes transparent, probably owing to crystallisation.

529. The rise in temperature when supercooled phosphorus is made to solidify may be shown by a slight modification of the experiment. A glass tube, 15 centimetres long and 7 millimetres bore (which is about as wide as can safely be used), and rather thin in the walls, is bent at right angles, at a point about 5 centimetres from the closed end. It is nearly filled with phosphorus in the manner described above. The tube is held in a clamp, with its shorter limb resting upon the face of a thermopile, which is connected to a galvanometer. When the phosphorus is solidified, as in the above experiment, the rise in

temperature will cause a marked deflection of the needle in the direction indicating heat.

530. **To show that phosphorus does not glow in pure oxygen.** A clean stick of phosphorus, suspended by a piece of string, is lowered down into a cylinder filled with oxygen, the room being as dark as possible. The glow which the phosphorus emits while exposed to the air will be instantly quenched by the oxygen.

531. **To show that phosphorus will glow in oxygen under reduced pressure.** A glass tube 30 centimetres long and $2\frac{1}{2}$ centimetres wide is drawn out at one end, and at the other fitted with a caoutchouc cork carrying a stopcock. The tube is connected by its small end to an air-pump, and the stopcock is connected to an oxygen supply. A stick of clean phosphorus is placed in the tube, and a stream of oxygen passed through. As in the above experiment, the phosphorus does not glow. The cock is now closed and the pressure reduced by one or two strokes of the pump, when instantly the glow makes its appearance. If the pressure be again restored by admitting more oxygen the glow once more ceases. As soon as the experiment is concluded the stick of phosphorus should be withdrawn, and replaced in water.

532. This experiment may be made on a larger scale, and with no risk whatever, by using, instead of a stick of phosphorus, a solution of the element in oil. The tube in this case may be much longer and broader, being bent, as shown in the figure, in order to contain the oil. The solution is made by placing a few pieces of phosphorus into a bottle of olive oil, and gently warming the mixture; in a short time the oil, if exposed to the air in the dark, will be seen to glow with a very considerable luminosity. A quantity of this solution is introduced into the tube by means of a bent funnel; the cork carrying a stopcock is then replaced and the tube exhausted, in which condition it may be allowed to remain until the experiment is actually to

be performed. The cock is then opened and the tube instantly filled with oxygen, when no glow will be seen, but on suddenly reducing the pressure the entire length of the tube instantly

FIG. 170.

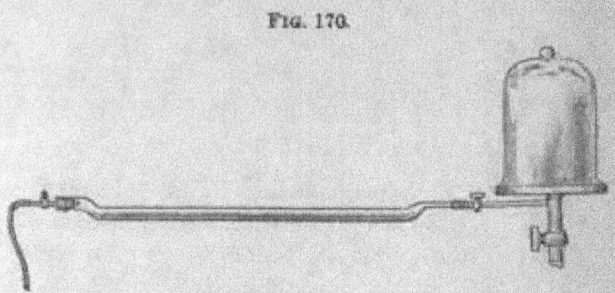

becomes luminous. In order to produce the effect suddenly it is well to connect the tube to an exhausted receiver, which is most simply done by placing an ordinary receiver upon the plate of the air-pump, and previously exhausting it.

533. To show that phosphorus will not glow in air under increased pressure. For this purpose a stout glass tube is sealed at one end and fitted at the other with a caoutchouc cork carrying a stopcock. The cork has cemented upon the top a brass plate with two projecting ears (fig. 171), and by means of two stout brass wires which are fastened to a wooden block, and

FIG. 171.

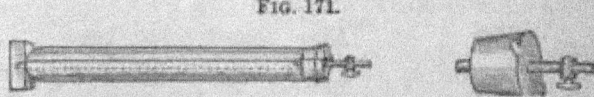

which at their free ends are bent into the form of a hook, the cork can be securely held in its position, so that it is not blown out when the pressure within the tube is increased.

A stick of phosphorus is placed in the tube and the cork secured in position. By means of a small condensing pump, air is forced into the tube, and as the pressure increases the glow will be seen to diminish until it altogether stops. On releasing the pressure by letting out the air the phosphorus once more begins to glow.

534. To show that certain vapours when present in air will prevent phosphorus from glowing. A few drops of ether, or turpentine, or carbon disulphide, are allowed to fall into a cylinder full of air, and the vapour to mix with the air. On introducing a stick of phosphorus, as in Exp. 530, the glow will be instantly stopped.

535. The action of phosphorus upon potassium chlorate may be shown by placing upon a block of wood a minute pinch of very finely powdered potassium chlorate, and moistening it with a drop of a solution of phosphorus in carbon disulphide, delivered from a fine pipette. In a few minutes, as the liquid evaporates, the mixture explodes with a loud report. This experiment should only be made upon a very small scale.

RED PHOSPHORUS

536. The formation of this allotropic modification, by the combustion of ordinary phosphorus in a limited supply of air, may be shown by igniting a piece of phosphorus contained in a dish floating upon water, and covering it with a bell-jar; as the combustion continues, and becomes more and more feeble owing to the diminished supply of oxygen, a quantity of red phosphorus is formed within the dish.

537. By the combustion of ordinary phosphorus in oxygen beneath warm water. A few pieces of phosphorus are placed in a glass cylinder which is nearly filled with hot water. A stream of oxygen gas is bubbled through the melted phosphorus, which burns at the bottom of the vessel, at the same time producing large quantities of the red modification.

538. To show the reconversion of red into yellow phosphorus. A small quantity of the red powder is placed in a test tube which is provided with a cork carrying two tubes. A stream of coal-

gas is passed through to displace the air, and while a very gentle stream is still circulating, the phosphorus is heated over

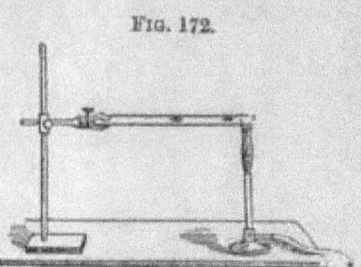

FIG. 172.

a Bunsen flame. The conversion readily takes place, and the ordinary phosphorus sublimes upon the upper parts of the tube in the form of clear yellow oily drops.

539. To compare the behaviour of the two forms of phosphorus towards heat. A strip of brass about 20 centimetres long and 2½ centimetres broad is held by one end in a clamp. At a distance of about 4 centimetres from the other end is placed a small quantity of red phosphorus, and, at a point about 10 centimetres farther on, a fragment of yellow phosphorus is placed. The end of the metal strip is heated by a Bunsen burner, which may be brought to within about 2 centimetres of the red modification. It will be seen that the distant piece of yellow phosphorus is the first to ignite.

540. A mixture of red phosphorus and potassium chlorate may be detonated by placing upon a smooth brick, or the bottom of an inverted mortar, a small pinch of finely-powdered potassium chlorate, and a rather smaller quantity of washed and dried red phosphorus. The two are gently mixed by means of a piece of paper. On stroking the mixture with the blade of a spatula, and thereby exposing it to a slight friction, it explodes or detonates with a sharp report.

541. A solid stick of potassium chlorate when drawn across the rubbing surface of a safety matchbox will produce a small flash of light; and if this be performed over a gas-lamp from which gas is escaping the gas will be inflamed. The piece of potassium chlorate may be obtained by fusing a quantity of the salt in a test-tube, and afterwards breaking away the glass.

542. If a lozenge, made of a mixture of potassium chlorate and sugar, be rubbed against the red phosphorus contained upon a safety matchbox, the lozenge will be inflamed, and will continue to burn. The ordinary chlorate of potash lozenge of the druggist has usually too little sugar in its composition to show this effect; but suitable pellets may be made by mixing one part, by weight, of finely-powdered white sugar with two parts of powdered potassium chlorate, moistening the mixture with sufficient water to make it into a stiff paste, and moulding it into little pellets between the fingers. When dry, one of these, when lightly rubbed against the matchbox, will at once ignite.

PHOSPHORETTED HYDROGEN. PH₃

543. Preparation by the action of sodium or potassium hydroxide upon phosphorus. $3NaHO + 4P + 3H_2O = 3NaH_2PO_2 + PH_3$. The gas evolved from this reaction being spontaneously inflammable, owing to the presence in it of a certain quantity of liquid hydride of phosphorus, it is necessary to displace the air from

FIG. 173.

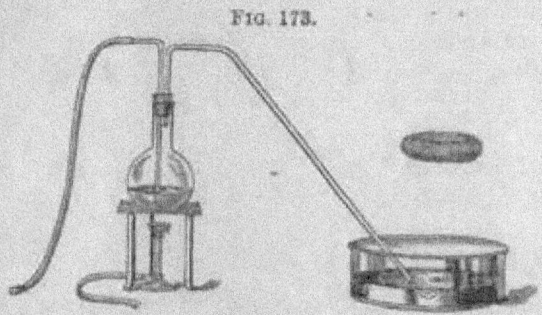

the apparatus used for generating the gas by some indifferent gas, such as hydrogen or coal-gas. For this purpose a flask is fitted with a cork carrying two tubes, as shown in the figure; one reaching nearly to the bottom of the flask, the other being a delivery tube. This latter tube is bent so as to enter a trough or dish placed at some distance from the flask, in order that the smoke rings which are produced in the experiment may rise quite clear of the air-currents produced by the Bunsen lamp. The flask is about one-third filled with strong caustic soda solution, and a few pieces of phosphorus introduced. A stream of coal-gas is then allowed to pass through the flask until the air is displaced, when the mixture may be gently heated by means of a rose burner. As the tube dips beneath the liquid contained in the flask, it is immaterial whether the caoutchouc tube used to convey the coal-gas be removed or not during the experiment, but as it is convenient to again connect with the gas supply at the close of the experiment, it may be allowed to remain, the gas being turned off. The delivery tube, which should be made of fairly wide bore tubing, is allowed to dip a short distance beneath the surface of water in a glass basin. As soon as the bubbles of gas which escape are spontaneously inflammable, the lamp should be lowered, or temporarily removed, in order to reduce the rate of evolution to such a condition that the bubbles are not delivered into the air faster than one in every four or five seconds. In this way, if the air be fairly quiet, fine vortex rings of smoke will be obtained, and at intervals sufficiently long to prevent them from interfering with each other, as would be the case if the gas were being generated too quickly.

544. **Preparation by the action of heat on sodium hypophosphite.** $2NaH_2PO_2 = HNaPO_4 + PH_3$. When hypophosphorous acid, or a hypophosphite, is gently heated, it evolves phosphoretted hydrogen and is converted into the ortho-acid, or a salt. The most convenient compound to employ is sodium hypophosphite (a commercial salt). A few grams of this salt are heated in a test tube provided with a cork carrying a delivery tube. Phosphoretted hydrogen is quickly evolved, and spontaneously ignites on coming into the air. This forms the readiest way of showing the formation and properties of phosphine, but as the evolution of the gas is not quite so much under control as when obtained from phosphorus and caustic

soda, it is possible that the smoke rings will not be so perfect. In this experiment it is not necessary to displace the air from the apparatus by coal-gas.

545. **Preparation by the action of alcoholic potash upon phosphorus.** Two or three sticks of caustic potash are introduced into a flask about one-third filled with methylated spirit, and a few fragments of phosphorus added. The gas produced in this reaction is practically free from the liquid hydride, and is therefore not spontaneously inflammable; it is not necessary, therefore, to fill the apparatus with coal-gas, so that the flask is provided only with a delivery tube conveying the gas to the pneumatic trough, where it may be collected in the usual way. Care must be taken not to heat the mixture too rapidly at first, for, although phosphoretted hydrogen is not spontaneously inflammable, its ignition point is extremely low, below the temperature of boiling water, so that it is quite possible to get an explosion in the flask while there is a mixture of air and phosphoretted hydrogen in it if the heat be carelessly applied. The flame of a small rose burner should be employed until the gas is being freely disengaged.

546. **Preparation by the action of caustic potash upon phosphonium iodide.** $PH_4I + KHO = KI + H_2O + PH_3$. A quantity of phosphonium iodide is placed in a small flask, fitted with a dropping-funnel and delivery tube. Strong aqueous potash is allowed very slowly to drop upon the compound, when an evolution of pure phosphine takes place.

Phosphonium iodide may be prepared by dissolving 30 grams of phosphorus in carbon disulphide in a small tubulated retort, fitted with a cork carrying a dropping-funnel and a short delivery tube; 50 grams of iodine are gradually added, the retort being kept cold during the operation by being placed in ice. The retort is then placed in a water-bath, and the carbon disulphide distilled off, a stream of carbon dioxide being passed through towards the close of the distillation to carry over the last portions. The condenser is then replaced by a straight wide glass tube, supported in a horizontal position, and into one end of which the neck of the retort is fixed. 18 cubic centimetres of water are poured into the dropping-funnel, and allowed to enter the retort by drops, a very slow stream of carbon dioxide being passed through the apparatus during the entire operation. An energetic reaction takes place, the heat of which volatilises the phosphonium iodide as it is produced, and it is carried along by the current of carbon dioxide into the wide tube, where it condenses. After all the water has been added, the retort may be gently heated, to drive off the remaining phosphonium iodide. It is advisable to cool the condensing tube by covering it with blotting-paper upon which ice-cold water is allowed slowly to drop.

547. The combustion of phosphoretted hydrogen in oxygen may be shown by means of the apparatus described for burning ammonia in oxygen (Exp. 278). The jet of phosphoretted hydrogen, delivered from a gas-holder, is first ignited in the air, where it burns with a brightly luminous flame ; when the flame is fed with oxygen its luminosity is increased to a dazzling intensity.

548. To show the action of bromine upon phosphoretted hydrogen. A drop of bromine allowed to fall into a test tube filled with phosphoretted hydrogen takes fire as it comes in contact with the gas.

549. To show the action of iodine upon phosphoretted hydrogen.

$$(1)\ PH_3 + 3I_2 = PI_3 + 3HI.$$
$$(2)\ PH_3 + HI = PH_4I.$$

A few grams of iodine are placed near to one end of a long horizontal glass tube about 20 millimetres diameter. A slow stream of phosphoretted hydrogen, dried by being passed through a chloride of calcium tube, is delivered into the tube at the end where the iodine is situated. The iodine is gently warmed by covering that part of the tube with a strip of filter paper, and allowing warm water to slowly drip upon it. At some little distance a portion of the tube is similarly cooled by means of ice-water. The action of the phosphoretted hydrogen

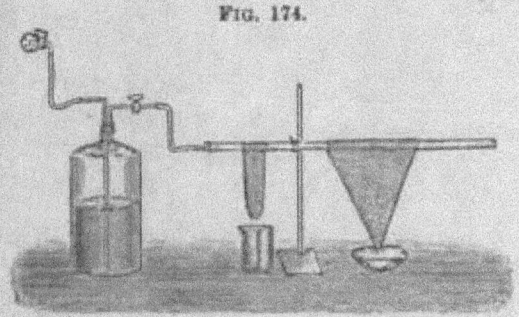

FIG. 174.

upon the iodine is in the first place to produce phosphorus tri-iodide and hydriodic acid, and the nascent hydriodic acid then combines with a further quantity of the phosphoretted hydrogen, forming phosphonium iodide, which condenses in the cooled part of the tube in the form of transparent crystals.

550. The direct combination of phosphoretted hydrogen and hydriodic acid may be shown by leading separate streams of the two gases into a cooled flask. For this purpose a flask is fitted with a cork carrying three tubes, two of which pass nearly to the bottom of the flask, while the third ends just below the cork; this exit tube is connected to a tube which dips beneath a shallow layer of mercury in the bottom of a beaker containing water. Hydriodic acid from a generating apparatus (see Experiment 240) is passed into the flask by one tube, and phosphoretted hydrogen (previously prepared and contained in a gas-holder) is delivered through the other. At ordinary temperatures only a very slight combination takes place, but if the flask be immersed in a freezing mixture of ice and salt, or ice and calcium chloride, dense white fumes at once appear, which condense upon the glass, forming a white solid.

551. When calcium phosphide is thrown into water a mixture of the gaseous and the liquid phosphoretted hydrogens is produced, which ignites spontaneously when it comes into the air. $2PCa + 4H_2O = 2CaH_2O_2 + P_2H_4$. In order to collect the liquid compound a quantity of calcium phosphide is placed in a flask fitted with a dropping funnel and two tubes. One of the tubes is connected to a carbonic acid generator, the other to a chloride of calcium tube, which in its turn is attached to a U tube immersed in a freezing mixture. A brisk current of carbon dioxide is passed through the apparatus to displace the air, after which a gentle stream only is allowed to pass. Water is gradually admitted by means of the dropping-funnel, when the action proceeds, and the liquid compound will collect in small quantities in the cooled U tube.

552. The action of calcium phosphide, as used in the 'Holmes signal lights,' may be illustrated by placing a few lumps of the phosphide in a short piece of glass tube about 35 millimetres diameter, drawn off in the form of a blunt cone at one end with an opening about 5 millimetres wide, and fitted at the other end with a cork carrying a short piece of smaller glass tube reaching about half-way up the tube, and ending just below the cork, as seen in the figure. The apparatus is fixed into a hole

Fig. 176.

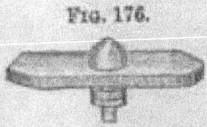

cut in a piece of wood in such a way that the conical end only just protrudes through the wood. The whole is then thrown upon the surface of water contained in a large trough, where it floats by virtue of the piece of wood. Water at once enters by the tube in the cork, and, coming into contact with the phosphide, generates phosphoretted hydrogen, which will spontaneously inflame at the top of the little apparatus and continue burning for some time.

PHOSPHORUS PENTOXIDE. P_2O_5

553. Preparation by the combustion of phosphorus in air. As a lecture experiment this is most conveniently done by burning a piece of phosphorus in a small porcelain dish standing on a

large slab of porcelain (such as is commonly used on the butterman's scales) or a sheet of plate glass; in the latter case a thin piece of cork should be placed under the small dish in which the phosphorus is burnt. The phosphorus is ignited, and a large bell-jar or glass shade, having a capacity of 15 or 20 litres, is placed over it. The phosphorus gradually burns itself out, and a layer of the oxide of a considerable thickness will collect on the slab of porcelain. On removing the glass shade the oxide may be collected together with a spatula, and put into a stoppered bottle.

FIG. 177.

554. The powerful affinity of phosphorus pentoxide for water may be demonstrated by throwing a quantity of the oxide into water, when the energy of combination will be manifested by the production of a hissing sound, resembling the quenching of a hot piece of metal by water.

555. If a small quantity of the pentoxide be placed upon a clock glass, and a piece of gun-cotton laid upon it, and a drop or two of water be delivered on to the oxide by means of a fine pipette, the heat developed by the combination of the oxide with the water will be sufficient to ignite the gun-cotton.

556. The power possessed by phosphorus pentoxide of abstracting the elements of water from certain compounds is seen in its action upon ethyl alcohol, whereby ethylene is produced. (See Ethylene, Experiment 455.)

PHOSPHOROUS OXIDE. P_4O_6

557. Preparation by the combustion of phosphorus under special conditions. Although this experiment is not very suitable as a lecture illustration, nevertheless, as it is the only method which has hitherto been devised for the formation of this com-

pound, a description of the process may be given. A condenser resembling an ordinary Liebig is constructed of brass tubes brazed together. The inner tube should be 60 centimetres long and 3 centimetres diameter, and should only project beyond the outer tube about 2 centimetres at each end, and one end of it should be slightly widened out. A piece of glass tube, the same diameter as the inner brass pipe and about 40 centimetres long, is bent as indicated in the figure, one end being slightly tapered that it may form a rough fit with the mouth of the brass tube. Both the tube and the condenser are supported by clamps. No luting is necessary where the glass fits into the metal, as the phosphorus pentoxide which is produced effectually closes any

FIG. 178.

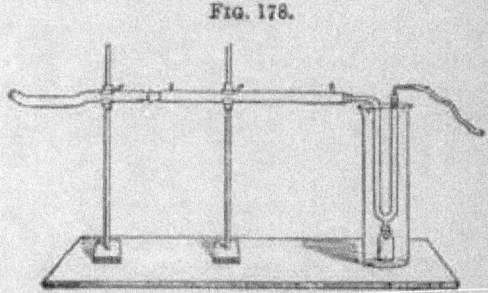

crevices. Into the other end of the condenser is fitted a cork, through which passes the bent limb of a U tube having a small outlet tube at the bottom, upon which is fitted a small bottle. Into that end of the condenser a short plug of glass wool is thrust, to arrest any of the pentoxide which is carried along the tube. Some little care is required to adjust this filtering plug to the best advantage; it should be about 8 or 10 centimetres long, and rather loosely packed, otherwise it soon stops up the tube altogether. The U tube with its bottle is immersed in a freezing mixture of ice and salt. A quantity of phosphorus is placed in the glass tube and ignited by introducing a hot wire, and a slow stream of air aspirated through the apparatus; as the tube becomes warmed by the burning phosphorus the rate at which air is drawn through may be increased, but the combustion must not be allowed to proceed too vigorously. As the phosphorus is consumed the point at which the combustion is taking place gradually creeps along the tube until it approaches near to the brass tube. The aspirator should then be stopped, and the glass tube removed and replaced by a second containing a similar charge of phosphorus, and the operation continued. At first the condenser should be allowed to remain cold, but when the operation has continued for about half an hour a stream of warm water, at about 60° or 70°C., may be made to circulate; the condensed P_4O_6 is thereby vaporised and carried on into the U tube, where it crystallises as a white solid. From time to time the freezing mixture may be removed, and the accumulated solid substance gently melted down into the little bottle by the application of warm water to the outside of the tube. For the weight of phosphorus consumed the yield of P_4O_6 is extremely small; after a little experience 40 to 50 grams may be obtained in a day.

558. One or two drops of the melted compound—which liquefies below the temperature of the hand—poured upon a piece of filter paper and introduced into a cylinder of chlorine, instantly inflames in the gas.

559. If a small flask filled with oxygen be immersed in a beaker containing boiling water, and a few drops of the melted compound upon filter paper be dropped into the warm gas, the substance combines with the oxygen with vivid combustion.

560 If dropped upon a hot and strong solution of caustic potash in a small dish it spontaneously ignites.

561. A small quantity of the oxide may be added to water in a test tube, when it is seen that it is almost insoluble in the cold, and only slowly dissolved upon warming.

SULPHUR

562. To show the formation of sulphur by the mutual action of sulphur dioxide and sulphuretted hydrogen.

$$5SO_2 + 5H_2S = H_2S_5O_6 + 4H_2O + 5S;$$

or

$$SO_2 + 2H_2S = 2H_2O + 3S.$$

Two cylinders of equal size, one containing sulphur dioxide and the other sulphuretted hydrogen, are brought mouth to mouth, and the covers removed. As the gases mix, a copious deposition of sulphur takes place upon the sides of the vessels. The cylinder of sulphuretted hydrogen should be collected over water, as the presence of the moisture promotes the reaction.

563. The action of heat upon sulphur may be shown by carefully melting a small quantity of pure sulphur in a Florence flask. Flowers of sulphur should not be used, but either good roll sulphur, or some that has been obtained by distillation. (See Experiment 566.) In order to make it easier to obtain the melted sulphur in the first or limpid condition, without over-heating any portion of it, it is well to previously melt it in the flask, and allow it to crystallise over the surface of the glass by revolving the vessel while it is cooling. In this way a thin layer of the element is obtained, which is very quickly and easily melted into a pale amber-coloured liquid, the limpidity of which may be shown by suddenly shaking the flask. On heating the flask further, the colour of the sulphur rapidly darkens, and if the flask be revolved as it is being heated, the sulphur is very quickly obtained in the second or viscid condition, when the flask may be inverted without the contents showing any tendency to run out. On still further heating the flask, until the sulphur begins to boil, the limpidity of the dark liquid mass may be shown by shaking the flask, or by pouring some of the contents out.

564. To prepare the prismatic variety of sulphur. A quantity of sulphur (either good roll or distilled) is melted in a beaker, the beaker being heated by means of a rose burner. By the exercise of a little care, and by stirring the mass from time to time with a glass rod, the sulphur may be melted without its becoming dark and viscous. When the whole is in the liquid condition, the lamp may be removed and the mass allowed to crystallise. When it has partially solidified, the still liquid portion is poured out into a plate, and on cutting away the crust upon the top the mass of prismatic crystals which line the interior of the beaker will be seen. The exact moment when the sulphur may be poured away is best judged by observing the formation of crystals upon the surface of the liquid. If the beaker be left to cool standing upon the wire gauze which supported it while it was being melted, the crystallisation will begin at the bottom, and, spreading up the sides, will extend across the surface. Before the surface has entirely congealed, the remaining liquid may be poured away. If this point be passed, and there is formed a complete crust upon the top, it must be broken through by means of a glass rod, in two places and the liquid then run out.

565. To prepare the plastic modification of sulphur. A quantity of broken roll sulphur is heated in a Florence flask

Fig. 179.

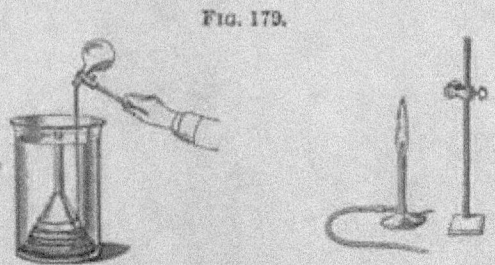

until it boils, when it is poured in a thin stream into cold water, where it congeals to the plastic condition. It is convenient to place, in the beaker of water into which the sulphur is poured, a large inverted funnel; by raising this at the end of the operation the mass of sulphur may be conveniently withdrawn from the water, and by pouring the stream of melted sulphur round and round the funnel the plastic thread may be more easily unwound.

566. A more beautiful result is obtained by distilling sulphur in a retort, and allowing the condensed liquid to fall into water. For this purpose a large non-tubulated retort is three parts filled with flowers of sulphur. The retort is supported in a clamp in such a position that its neck forms a rather steep incline, and it is heated by means of a large Bunsen lamp. For the success of the experiment it is necessary that the sulphur be distilled with such rapidity that the liquid falls from the end of the retort in a continuous stream, and not in drops; it is therefore well to cover the top of the retort with a piece of asbestos cloth (fig. 180). As soon as the sulphur has fairly begun to distil, a lighted taper is applied to the end of the retort, and the uncondensed vapour inflamed. The liquid then falls into the water, and, owing to an undulatory movement that it assumes, the sulphur as it condenses coils itself up in the most regular and beautiful manner, forming spirals of plastic sulphur of a delicate translucent amber colour.

FIG. 180.

If the precaution of igniting the issuing sulphur be omitted, the effect will be entirely spoilt, for the vapour which issues from the retort along with the liquid will condense as a film or skin upon the surface of the water, and prevent the stream of liquid from passing in, but the sulphur will run over the surface of the water until a quantity has collected, which then suddenly falls through the film as a large irregular blob, after which a fresh quantity again collects on the surface, and so on.

The distillation should be continued until the whole has passed over, for if a quantity of sulphur be left in the retort and allowed to solidify, it is extremely likely to fracture the glass when it is again heated.

567. The combustion of certain metals in sulphur vapour may be shown in the following way:—A quantity of flowers of sulphur is heated in a Florence flask until it boils, and the vapour burns at the mouth of the flask; a spiral of copper ribbon, obtained by rolling copper wire, is lowered through the flame into the vapour, when it burns with considerable energy. A bundle of steel wire, heated to bright redness in a blowpipe flame and introduced into the flask, burns in the sulphur vapour, the molten sulphide dropping into the sulphur.

The combination of sulphur and iron may also be shown in a striking manner by pressing a roll of sulphur against the end of a red-hot bar of iron. A thick bar of iron is made as hot as possible in a furnace, the nearer the heat approaches to whiteness the better; it is withdrawn from the furnace, and a stick of sulphur pressed against it. The longest roll of sulphur which can be obtained should be used. The iron combines with the sulphur, and the molten sulphide falls to the ground in a brilliant shower.

568. The spontaneous oxidation of moist sulphur may be shown by stirring flowers of sulphur and water together for a few minutes, and throwing the mixture on to a filter. The clear filtrate will redden litmus paper, and give the test for sulphuric acid with barium chloride.

569. When sulphur and potassium chlorate are rubbed together the mixture explodes with considerable violence. To show this a minute quantity of potassium chlorate, not more than 1 gram, is placed in a mortar, and about the same quantity of sulphur added. The mixture is rubbed with the pestle, when a sharp detonation will result.

SULPHURETTED HYDROGEN. H_2S

570. Preparation by the action of sulphuric acid upon ferrous sulphide. $FeS + H_2SO_4 = FeSO_4 + H_2S$. A quantity of broken ferrous sulphide is placed in a two-necked Woulf's bottle fitted with a thistle funnel and delivery tube. Water is introduced to a depth of about two or three centimetres, and on adding a few drops of strong sulphuric acid a rapid evolution of the gas takes place. The gas may be washed by passing through a second smaller Woulf's bottle half filled with water, and finally collected at the pneumatic trough over water as warm as can be comfortably manipulated. The pneumatic trough should be either glass or earthenware.

571. Preparation by the action of hydrochloric acid upon antimonious sulphide. $Sb_2S_3 + 6HCl = 2Sb_2Cl_3 + 3SH_2$. Black antimonious sulphide is heated with strong hydrochloric acid in a flask fitted with a delivery tube, and the gas collected as above.

572. The synthetical formation of sulphuretted hydrogen may readily be demonstrated by passing a slow stream of hydrogen over a quantity of heated sulphur. For this purpose a few fragments of sulphur are placed in a combustion bulb and hydrogen slowly passed through the bulb tube. On heating the sulphur until it begins to vaporise, appreciable quantities of sulphuretted hydrogen will be evolved, and if the issuing gas be caused to impinge upon paper which has been impregnated with a lead salt, the presence of the sulphuretted hydrogen will be manifest by the rapid blackening of the test paper.

The same result will be obtained by substituting iron pyrites for the sulphur, illustrating the mode of formation of sulphuretted hydrogen in coal-gas.

573. The combustion of sulphuretted hydrogen may be shown by igniting a jet of the gas as it issues from the generating vessel. If a glass rod moistened with ammonia be held above the flame, the formation of white fumes will demonstrate the presence of sulphurous acid in the products of combustion. By holding a clean dry glass cylinder over the flame the water from the combustion will be deposited upon the glass. This may be shown to be acid by means of litmus; it may also be shown to contain sulphuric acid by rinsing the cylinder with a small quantity of distilled water and applying the barium chloride test.

574. If a cylinder of sulphuretted hydrogen be inflamed at the mouth, it will be seen, as the combustion proceeds down the cylinder, that a quantity of sulphur is deposited upon the interior of the vessel. This deposition of sulphur, by the incomplete combustion of sulphuretted hydrogen, may also be seen by depressing a piece of cold glass upon the flame of the burning gas as it issues from a jet.

575. The incomplete combustion of sulphuretted hydrogen may also be secured by causing a jet of air to burn in an atmosphere of sulphuretted hydro-gen. For this purpose a piece of apparatus similar to that already described for the combustion of air in coal-gas may be used. In this case, however, the air must be de-livered from a gas-holder, as sul-phuretted hydrogen, not being lighter than air, will not ascend in the ap-paratus and draw air after it. A stream of sulphuretted hydrogen is delivered into the lamp glass by the small tube (A, fig. 181), and inflamed as it issues from the top; the tube T,

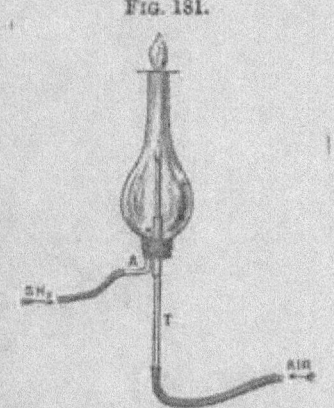

FIG. 181.

through which air is delivered, is passed up through the short wide tube, the former being of such a size that it will easily slide up and down in the wide tube. This air tube is raised until it just reaches into the flame of burning gas at the top, when the air is slowly turned on; the jet of air will ignite

inside the flame of sulphuretted hydrogen, and the tube may
then be drawn down into the chimney, where it will continue
to burn.   Its combustion is attended with the separation of
considerable quantities of sulphur, which deposits upon the
interior of the glass chimney.

576. A mixture of two volumes of sulphuretted hydrogen
and three volumes of oxygen may be exploded in a soda-water
bottle on the application of a lighted taper.

577. The explosion of a mixture of sulphuretted hydrogen
and oxygen may be brought about by means of oxide of iron.
The mixture, in the proportion as above, is contained in a short
stout glass cylinder, and a quantity of ferric oxide introduced
by dipping into the cylinder a piece of tow attached to the end
of a wire, and which has been sprinkled with the oxide.   The
rapid absorption of the sulphuretted hydrogen by the oxide will
in a few seconds develop sufficient heat to ignite the remaining
mixture of gases.   All specimens of oxide of iron are not equally
effectual in causing the result; a number of samples must be
tried, and that one selected and set aside which gives the best
result.

578. If a mixture of sulphuretted hydrogen and air be
passed over ferric oxide, the oxide becomes incandescent, owing
to the rapid sulphurisation and reoxidation of the compound.
This may be shown by allowing a stream of sulphuretted
hydrogen, issuing from a fine jet, to impinge upon a little
heap of ferric oxide placed upon a piece of wood, the jet being
held at a short distance from the oxide, so that a certain amount
of air is drawn along with it.   Under these circumstances the
oxide will be seen to glow.

The experiment is still more strikingly shown by placing the
ferric oxide in a moderately wide piece of glass tube, into one
end of which the chimney of a Bunsen lamp is fixed by means
of a piece of caoutchouc tube, as shown in fig. 181A.

A stream of sulphuretted hydrogen is admitted into the lamp, and thereby becomes mixed with air, after the usual

Fig. 181A.

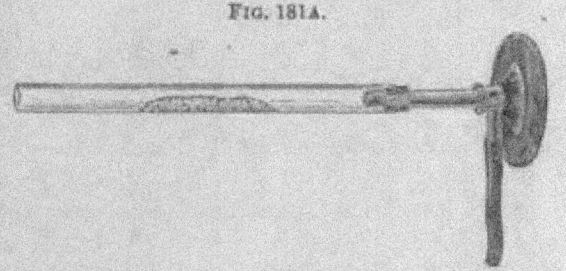

manner of the Bunsen. As the mixture of sulphuretted hydrogen and air passes over the ferric oxide, the latter almost immediately becomes incandescent. By conducting the experiment in this way the gases which issue from the tube can either be burnt, or conveyed away to a draught.

As the amount of light emitted by the glowing ferric oxide is rather feeble, it is advisable to darken the room during the experiment. Under these conditions the incandescence of the material will be manifest at a distance.

579. The action of chlorine upon sulphuretted hydrogen may be shown by inverting a cylinder of the latter gas upon a similar one containing chlorine, and withdrawing the plates. As the gases commingle the sulphuretted hydrogen is decomposed, sulphur being deposited upon the glass, and hydrochloric acid being formed. The sulphuretted hydrogen should be collected over water, in order that the gas in the cylinder may be moist.

580. If a jet from which sulphuretted hydrogen is issuing be lowered into a jar of 'euchlorine' (the gas obtained by the action of hydrochloric acid upon potassium chlorate), the sulphuretted hydrogen will spontaneously inflame. Euchlorine, being a mixture of chlorine with chlorine oxide, is liable to vary in composition, so that sometimes the sulphuretted hydrogen is not so readily inflammable as at others. If the gas, however, be first inflamed at the jet, so as to make the end of the glass warm, and then blown out and introduced into the euchlorine, it will spontaneously ignite without fail.

581. The oxidation of sulphuretted hydrogen by means of charcoal. (See Carbon, Experiments 394 and 395.)

582. The action of sulphuretted hydrogen upon solutions of metallic salts may be shown by allowing a stream of gas to pass through a series of cylinders containing suitable salts in solution. Four or five illustrations may be given, and should be selected so as to yield a variety of coloured precipitates. The following is a typical series. Copper sulphate; cadmium chloride; antimonious chloride; lead nitrate; zinc sulphate. The solutions should be very dilute, so that it may not take too long to complete the precipitations. The antimonious oxychloride, which is precipitated on diluting the chloride, may be dissolved by throwing in a crystal or two of tartaric acid, and the zinc sulphate must be made alkaline with ammonia. The sulphuretted hydrogen is generated in a Woulf's bottle, the thistle funnel in which is sufficiently long to enable sufficient pressure to be maintained in the bottle to overcome the resistance of the entire series through which the gas has to be passed.

FIG. 182.

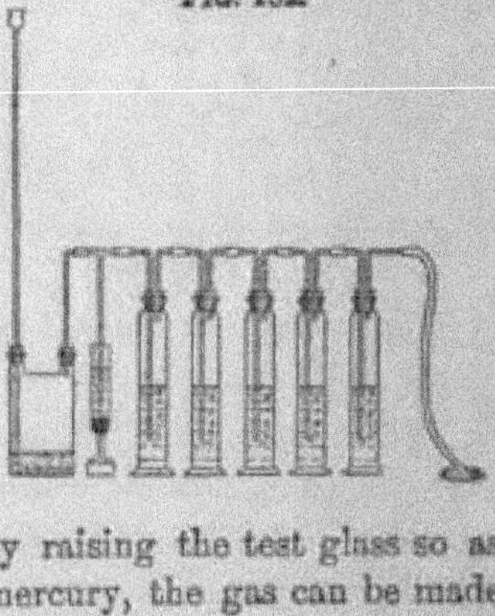

It is convenient to insert a T piece between the generating bottle and the series of cylinders, which may be arranged to dip into a test glass containing sodium hydrate, having a quantity of mercury in the bottom; by raising the test glass so as to cause the tube to dip into the mercury, the gas can be made to pass into the metallic solutions at will.

HYDROGEN PERSULPHIDE. H_2S_2

583. Preparation by the action of hydrochloric acid upon calcium persulphide. $CaS_5 + 2HCl = CaCl_2 + 3S + H_2S_2$. A large beaker is three-parts filled with hydrochloric acid diluted with twice its volume of water, and a slow stream of calcium persulphide is poured into it, with constant stirring, care being taken to keep the acid in excess. Hydrogen persulphide separates out in the form of minute oily drops, which gradually fall and collect at the bottom as a yellow oily liquid. A quantity of sulphur is at the same time precipitated.

584. The action of certain peroxides upon hydrogen persulphide may be shown by adding a small quantity of manganese dioxide to some of the persulphide in a test tube. It is an advantage to dissolve the thick oily liquid in about its own volume of carbon disulphide, as such a solution allows the manganese dioxide more easily to mix with it, and also may be longer preserved than the undiluted compound.

585. If a few drops of the solution be added to water which has been tinted with litmus, and the mixture shaken together, the colour will be discharged.

SULPHUR DIOXIDE. SO_2

586. Preparation by the combustion of sulphur in oxygen. A quantity of sulphur is heated in a combustion bulb, and a stream of oxygen passed over it. The products of the combustion should be passed through a tube packed with glass wool, and the dioxide may then be collected by displacement.

587. Preparation by the action of sulphuric acid upon copper. $Cu + 2H_2SO_4 = CuSO_4 + 2H_2O + SO_2$. A quantity of thin copper clippings are placed in a capacious flask and covered with strong sulphuric acid, and the mixture carefully heated. The flask is fitted with a mercury safety-funnel, and the gas, after

being dried by its passage through a Woulf's bottle containing
sulphuric acid, is collected over mercury.   (See fig. 183.)

FIG. 183.

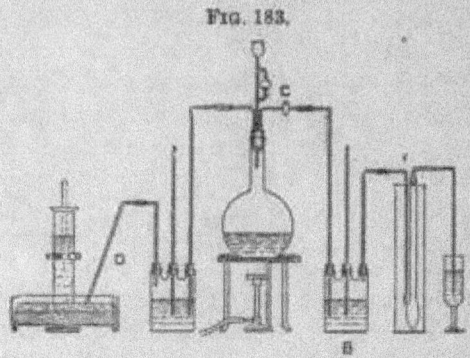

**588. Preparation by heating a mixture of manganese dioxide
and sulphur.** $2MnO_2 + 3S = 2MnOS + SO_2$. A mixture, in the
proportions of about 3 parts of sulphur to 4 of the dioxide,
is heated in a short piece of combustion tube, closed at one end,
and provided with a cork and delivery tube.

589. To prepare liquid sulphur dioxide. This is most conveniently done by passing the gas obtained by the action of
sulphuric acid upon copper in Experiment 587 through a
condensing tube immersed in a freezing mixture of chloride
of calcium and ice. The apparatus may be arranged as shown
in fig. 183. By closing the stopcock c the gas is compelled to
pass to the mercury trough, and may be there collected; but
when c is open, and the delivery tube D is kept beneath the
mercury, the gas will escape through the bottle B, being the
way of least resistance, and will be condensed in the tube I
in the freezing mixture. Both the Woulf's bottles contain
sulphuric acid in order to dry the gas.

590. For the liquefaction of sulphur dioxide by pressure
alone, and also by pressure and cold, in the Faraday tube, see
Liquefaction of Gases, Experiments 646–648.

591. The solubility of sulphur dioxide may be shown by opening a cylinder of the gas, collected over mercury, beneath the surface of water; as soon as the plate is withdrawn the water rapidly fills the cylinder. Although the absorption is not so violent as in the case of such very soluble gases as ammonia, it is nevertheless advisable to employ a thick glass plate as a cover to the cylinder. In no case should a stoppered vessel be used.

592. The solid hydrate of sulphurous acid (H_2SO_3, $8H_2O$) may be obtained by passing a stream of sulphur dioxide into water cooled to $0°C$., or, better, to a few degrees below; as the water becomes saturated the crystalline compound separates out, and the mass becomes semi-solid.

593. To show the bleaching action of sulphur dioxide. For this purpose an infusion of red rose-leaves is made by pouring boiling water upon a quantity of the dried leaves (to be obtained from any druggist), and acidifying the infusion by the addition of a few drops of sulphuric acid. Upon allowing the gas to bubble through the solution, or, more quickly, by adding some freshly prepared and strong solution of the gas in water, the red colour of the infusion will be discharged, or changed into a pale straw colour. The original colour may be restored by the addition of a further quantity of strong sulphuric acid.

594. The composition of sulphur dioxide may be demonstrated by burning a fragment of sulphur in a measured volume of oxygen.

The apparatus described under Carbon Dioxide (Experiment 484) may be employed for this purpose.; in this case the small fragment of sulphur may be fused on to the platinum wire.

595. The decomposition of sulphur dioxide by the action of light may be shown by filling a long cylinder with the gas, collected by displacement, and covering the mouth with a plate glass by means of resin cerate. A parallel beam of electric light is passed through the cylinder, and almost immediately the path

of the beam will become more and more visible, owing to the formation of a gradually thickening cloud or fog within the cylinder, consisting of sulphur and sulphur trioxide. The cylinder should be sufficiently wide to clear the beam—that is to say, the

FIG. 184.

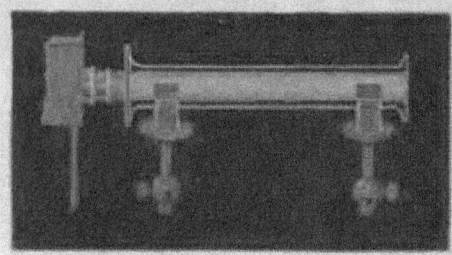

light should not strike upon the edge of the cylinder—or the resin cerate will be softened by the heat, and the plate be liable to slip from its place. When the cylinder is removed from the light the contents gradually become once more clear and transparent.

596. **To show the action of sulphur dioxide upon peroxide of lead.** A quantity of the peroxide is placed in a piece of combustion tube and slightly warmed by brushing a Bunsen flame along the tube. On passing a stream of sulphur dioxide over the mass, the substances combine with so much energy that the peroxide becomes red hot.

597. **To show the action of sulphur dioxide upon sodium peroxide.** A small quantity of sodium peroxide is dropped into a cylinder of sulphur dioxide. Each particle of the peroxide as it falls into the gas unites with it, the union being attended by brilliant combustion.

598. The combustion of iron in sulphur dioxide may be demonstrated by passing the gas through a piece of combustion tube in which is placed a quantity of reduced iron. Upon gently heating the metal it will be seen to burn in the stream of sulphur dioxide.

599. The low temperature produced by the spontaneous evaporation of liquid sulphur dioxide may be demonstrated by pouring a few cubic centimetres of the liquefied gas into a small beaker standing upon a block of wood, upon which a few drops of water have been placed; almost immediately the beaker will be frozen to the block.

600. If a quantity of liquid sulphur dioxide be thrown upon a few cubic centimetres of water in a dish, the water will be at once frozen to a solid mass, consisting of ice mixed with the solid hydrate of sulphurous acid.

601. This experiment may be so modified as to illustrate the phenomenon of freezing water in a red-hot vessel.

A platinum dish is strongly heated by means of a powerful Bunsen flame, and into the red-hot vessel 15 cubic centimetres of liquid sulphur dioxide are thrown. The liquid instantly assumes the spheroidal state, and while in this condition its temperature is practically that of boiling sulphur dioxide, viz. $-7°C$. Directly this liquid has been thrown into the hot dish, 15 cubic centimetres of water, contained in a test tube, are thrown upon it; the mass at once solidifies, and should be instantly thrown out of the dish on to a plate. The three operations—viz. throwing in the liquid dioxide, adding the water, and tipping out the contents of the dish—must be performed in rapid succession.

SULPHUR TRIOXIDE. SO_3

602. Formation by the direct union of sulphur dioxide and oxygen. For this purpose a stream of the two gases is passed over platinised asbestos contained in a combustion bulb. The two gases are delivered into the same Woulf's bottle, where, by causing them to bubble through sulphuric acid, the rate at which they are passing can be seen. The sulphur dioxide may be obtained by generating it from copper and sulphuric acid, or, more conveniently, from a supply of the liquefied

gas, which may be procured as an article of commerce in glass
soda-water 'syphons.'

So long as the platinised asbestos is cold no combination of

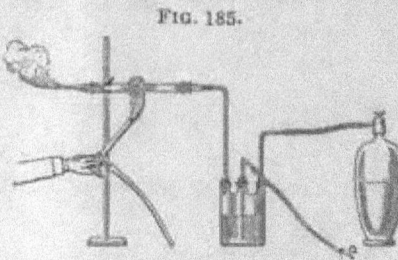

FIG. 185.

the two gases takes place,
but on gently heating the
bulb white fumes of the trioxide will be seen to issue
from the end of the tube.
If desired the trioxide may
be condensed; in which
case the operation should
be performed on a larger
scale, and the product passed through a U tube immersed in
a freezing mixture.

**603. The preparation of the trioxide by the distillation of
Nordhausen acid.** $H_2S_2O_7 = H_2SO_4 + SO_3$. A quantity of the
Nordhausen acid is gently heated in a retort, and the distillate
collected in a receiver, which is cooled by being placed in a
freezing mixture. If it be desired to preserve the compound as
a specimen for exhibition, a small quantity of the distillate may
be received in a round flask, with its neck previously drawn
out; the neck is then sealed before the blowpipe, and the
contents of the flask may be very slowly sublimed.

604. To show the action of water upon sulphur trioxide.
A small quantity of the solid is dropped into water, when combination takes place with a violent hissing sound.

605. To show the action of sulphur upon sulphur trioxide.
A small quantity of flowers of sulphur is added to the trioxide
contained in a small flask; a deep-blue compound is at once
formed (supposed to be the sesquioxide, S_2O_3), which, as the
temperature rises, rapidly decomposes with the evolution of
sulphur dioxide.

SULPHURIC ACID. H$_2$SO$_4$

606. Formation by the direct union of sulphur dioxide and hydrogen peroxide. A few cubic centimetres of hydrogen peroxide are poured into a cylinder of sulphur dioxide. The gas is rapidly absorbed, the liquid becoming warm by the combination of the two substances, and sulphuric acid is produced.

607. The manufacture of sulphuric acid, and the formation of the white crystalline compound, may be illustrated on a small scale by means of the apparatus shown in fig. 186. A large round flask or bolthead F, of about six or eight litres capacity, is fitted with a cork carrying five tubes, four of which reach to

FIG. 186.

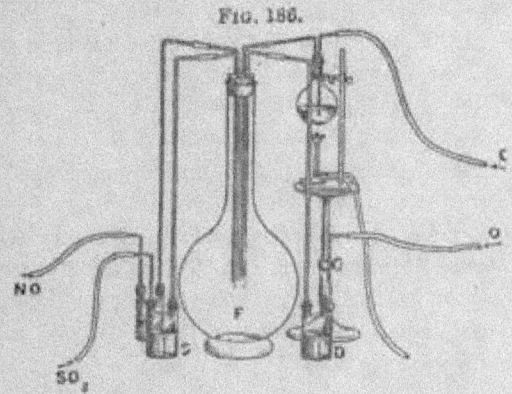

about the middle of the flask, the other, which passes just through the cork, serving as an exit tube. Three of the tubes are connected to Woulf's bottles containing sulphuric acid, through which the gases, oxygen, nitric oxide, and sulphur dioxide, are separately passed. The nitric oxide should be previously prepared, and stored in a gas-holder. The sulphur dioxide may be conveniently obtained from a siphon of the liquefied gas, and the oxygen from a cylinder of compressed gas. The fourth tube is connected with a small flask containing water, which can be heated, and through which oxygen gas can be made to bubble, so that in this way either dry

T

oxygen or oxygen charged with aqueous vapour can be passed into the vessel at will. In the first place a rapid stream of oxygen is delivered into the flask through the Woulf's bottle D, in order to replace the air within the vessel by oxygen. Nitric oxide is then gradually allowed to enter, which at once generates the brown oxides of nitrogen. When the nitric oxide is entering, a stream of sulphur dioxide is bubbled through the bottle S, at about the same rate as the nitric oxide is passing, and when this has continued for a little time, a small quantity of oxygen is introduced through the hot water to moisten the gases. In a few moments crystals begin to form all over the interior of the flask, and by regulating the supply of the different gases the deposit of the crystalline compound can readily be made to cover the surface of the entire vessel. When sufficient of the compound has been produced, the gaseous contents of the flask should be swept out, by a rapid stream of oxygen being admitted through D, until the atmosphere within the flask is quite colourless; and then the decomposition of the crystals by means of water and the re-generation of nitroxygen fumes may be shown, either by passing steam into the vessel from the flask E, or by removing the cork and pouring in a small quantity of water. In either case a rapid solution of the compound takes place, with the evolution of large quantities of nitroxygen fumes.

608. **To show the contraction in volume upon mixing sulphuric acid and water.** A long piece of combustion tube is closed at one end, and roughly graduated into three equal divisions by india-rubber rings. Strong sulphuric acid is introduced up to the second ring, and water is added up to the third, which should be about 10 centimetres from the top of the tube. The tube is then corked and the contents mixed by inverting the tube a few times, when, notwithstanding the great rise of temperature, the volume will be seen to have undergone a very appreciable contraction.

609. The heat developed on adding water to sulphuric acid

may be shown in the above experiment by moistening the outside of the tube with cold water, which will be vaporised by the heated mixture with so much rapidity that it will be seen to steam.

If the acid and water be mixed in a beaker, and a small flask containing a little ether, and fitted with a cork and straight exit tube, be dipped into the hot mixture, the ether will be caused to boil rapidly, so that the issuing vapour can be inflamed.

610. The powerful affinity of sulphuric acid for water may be demonstrated by its action upon such organic compounds as formates, oxalic acid (see Carbon Monoxide), or upon sugar.

120 grams of lump sugar are placed in a large beaker, and 90 cubic centimetres of warm water poured upon it, and the mixture covered and allowed to stand. In a short time the sugar will have entirely dissolved. The beaker is stood on a plate, and 120 cubic centimetres of strong sulphuric acid are quickly poured into the syrup, when the liquid blackens and instantly froths up over the top of the beaker as a spongy mass of charcoal, clouds of steam at the same time being evolved.

611. The power of sulphuric acid to char organic matter may be strikingly shown by brushing a little dilute acid upon paper (a word may be written, either with a camel's-hair brush or by using the finger), and then gently warming the paper by holding it some distance above a rose burner or in front of a fire. As the dilute acid becomes concentrated by evaporation, the paper is blackened where it was previously wetted.

This method may be made use of for showing the presence of sulphuric acid in the products of combustion of ordinary coal gas. A large flask nearly full of cold water is held by its neck in a clamp upon a retort-stand, and heated below by means of a rose burner. The flask must not be stood upon a ring or piece of gauze. Water at once begins to collect upon the outside of the flask, and this water gradually becomes acid owing to the combustion of sulphur compounds in the gas. As the flask and the water inside it get warm the condensed water outside begins to evaporate, and the traces of sulphuric acid it contains become more and more concentrated. After the gas has been burning for about ten minutes (the time depending upon the quantity of sulphur impurities the gas contains), the flask is placed upon a piece of white paper so as to make a wet ring upon the paper with the slightly acid water upon the outside of the flask. The paper is then gently warmed over a small rose burner, when the paper will become charred where it was wetted.

CARBON DISULPHIDE. CS₂

612. **Preparation by the direct union of sulphur and carbon.** The process by which carbon disulphide is manufactured may be illustrated upon a small scale by heating a porcelain tube, filled with fragments of charcoal, in a furnace placed in an inclined position. The upper end of the tube is closed with a cork, while the lower end carries a glass tube which is made to dip beneath the surface of water in a beaker. The tube is heated to redness throughout as long a length of it as convenient, the temperature being carefully regulated, as it will be found that the yield of the product rapidly diminishes as the temperature approaches to that of bright redness. The cork in the top of the tube is from time to time removed, and a fragment of sulphur introduced. The sulphur melts and runs down the tube, and the vapour coming in contact with the hot charcoal combines with it, and the disulphide, contaminated with more or less free sulphur, will collect at the bottom of the water in the beaker.

613. The combustion of carbon disulphide in air and in oxygen may be shown by inflaming a small quantity of the liquid contained in a porcelain dish, and causing a stream of oxygen to impinge upon the surface of the burning liquid.

614. A more striking method consists in boiling the liquid

in a flask fitted with a cork, through which passes a narrow brass tube. This brass tube has brazed upon it a wider tube, with a small branch tube attached, as shown in the figure. On applying a gentle heat to the flask a stream of vapour issues from the brass tube, and may be inflamed. By brushing a Bunsen flame once or twice over the metal any condensation of the disulphide in the tube will be prevented, and the flame will be perfectly steady.

FIG. 187.

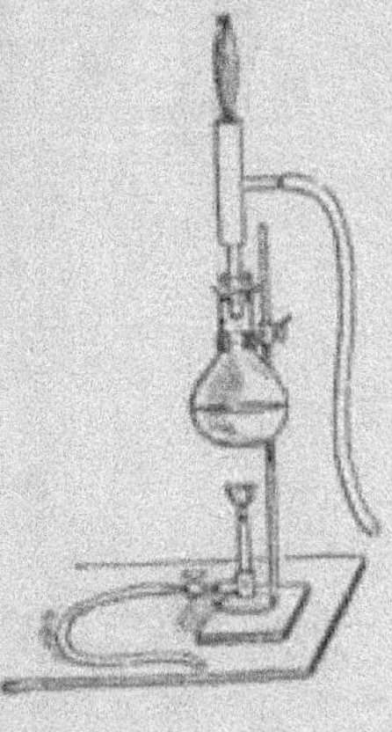

On supplying the flame with oxygen, by sending a stream of that gas through the outer tube, a dazzling flame is produced.

615. To explode a mixture of oxygen and carbon disulphide. A stout glass tube sealed at one end, about 20 centimetres long and 18 or 20 millimetres bore, is filled with oxygen, and two or three drops of carbon disulphide introduced; a cork is inserted, and a moment or two allowed for the vapour to mix with the gas. On applying a light to the mouth of the tube the mixture explodes with some violence.

FIG. 188.

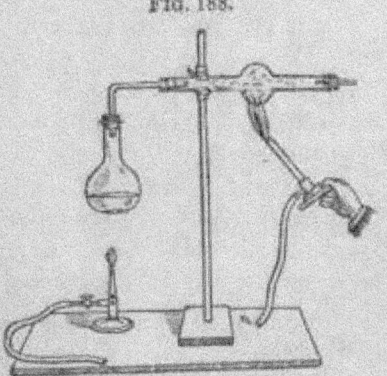

616. To show the combustion of potassium in carbon disulphide vapour. Carbon disulphide contained in a small flask is gently warmed, and the vapour passed through a combustion bulb in which is placed a fragment of potassium. On heating the metal it readily takes fire, and burns with considerable energy in the vapour.

617. If a bundle of steel wires be introduced into the flame of carbon disulphide burning from a jet, the iron burns with brilliant scintillations. For this purpose carbon disulphide is heated in a flask carrying a tube bent at right angles and drawn to a point; the issuing vapour is inflamed, and the bundle of wires held in the flame near to the tip.

FIG. 189.

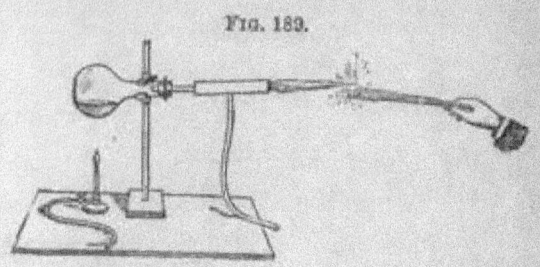

This experiment is rendered more certain if the flame of disulphide be fed with a little oxygen, which may be best accomplished by using the apparatus described in Experiment 614. The flask should be supported in an inclined position, so that the falling molten sulphide of iron may not drop upon the glass.

618. To show the action of calcium hydrate upon carbon disulphide. $CS_2 + CaH_2O_2 = 2CaO + CO_2 + 2SH_2$. Coal-gas is made to pass through a short glass tube which contains a small quantity of cotton-wool moistened with carbon disulphide. The gas so charged with the vapour of the disulphide is passed over a quantity of slaked lime contained in a combustion bulb; so long as the bulb is cold no action takes place, but on gently warming the lime there is a copious evolution of sulphuretted hydrogen, which may be proved by applying a moistened paper impregnated with acetate of lead, which will be instantly blackened.

Sometimes the coal-gas itself will be found to contain sufficient carbon disulphide to show this reaction; but the impurity is usually so carefully removed from the gas that if this experiment be made with a view to showing the method for removing the compound from coal-gas, an additional quantity must be thrown into it as described.

619. The decomposition of carbon disulphide by the detonation of mercuric fulminate. When a small quantity of fulminate of mercury is caused to detonate in the vapour of carbon disulphide, the disulphide is decomposed, and an instantaneous deposition of carbon and sulphide of mercury takes place.

The experiment is most conveniently performed in the following way:—A stout glass tube, 60 centimetres long and 2 centimetres bore, is fitted at one end with a cork through which pass two wires; one of these ends in a small capsule or deflagrating spoon, while the other is bent so that the end points into the capsule in such a way that an electric spark made to pass between them will discharge into the capsule (see fig. 190).

Fig. 190.

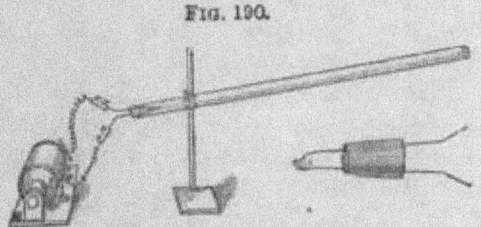

A small quantity of fulminate of mercury (not more than ·05 gram) is placed in the capsule, and the cork introduced into the lower end of the tube, which is held in a slightly inclined position, as shown in the figure. A long spill of blotting-paper, reaching nearly the whole length of the tube, is introduced at the open end, and by means of a fine-drawn-out pipette a quantity of the disulphide is allowed to trickle down the paper until it is nearly saturated with the liquid. Care must be taken to avoid letting the liquid run down to the cork, or come into contact with the fulminate. In a few moments the tube will be filled with the heavy vapour of carbon disulphide; the paper is then withdrawn, and the fulminate fired by an electric spark.

A sharp crack follows the passage of the spark, and the tube is instantly coated throughout its entire length with a black deposit consisting of carbon and sulphide of mercury.

It is supposed by some that this decomposition of the disulphide of carbon is caused by the actual shock produced by the detonation of the fulminate; that, in other words, the particular vibration set up by the explosion is the cause of the rupture of the molecules of the disulphide. It is quite conceivable, however, that the chemical affinity between the mercury, which is evolved from the exploding fulminate in a state of vapour at a high temperature, and the sulphur in the disulphide is in a measure concerned in effecting the decomposition, if, indeed, it is not the chief cause; it is significant that other explosives which do not contain a metal with which the sulphur could combine appear quite incapable of bringing about the same result.

Mercuric fulminate is most readily made in the following way:—10 grams of mercury are dissolved in 80 cubic centimetres of strong nitric acid (sp. gr. 1·42) with gentle warming. When the whole is dissolved 80 cubic centimetres of water are added and the solution cooled; 100 cubic centimetres of methylated spirit are added to the solution, and the mixture gently warmed in a flask of about 1 litre capacity. A brisk action soon sets up, which, however, with these proportions, never becomes uncontrollable, and the fulminate of mercury separates out as a greyish-white crystalline substance. The liquid is decanted off and the crystals washed and drained. The dry fulminate should be kept in bottles containing only a small quantity in each, and the bottles should be closed with corks and not with glass stoppers.

ARSENIC

620. The element may be prepared on a small scale by reducing arsenious oxide. A mixture of arsenious oxide, sodium carbonate, and powdered charcoal is introduced into a piece of combustion tube closed at one end, and heated by means of a Bunsen flame; the arsenic sublimes, and condenses as a black lustrous deposit upon the tube. Care should be taken to thoroughly dry the charcoal and the sodium carbonate before using them for this experiment, as the moisture they contain, condensing upon the tube, is liable to run back and cause its fracture, and also to prevent the formation of a good mirror.

621. The volatilisation of arsenic in a current of hydrogen may be shown by gently heating a fragment of the element in a piece of rather wide combustion tube, through which a slow stream of hydrogen is passed; the issuing gas should be allowed to escape into a suitable draught flue. As the piece of arsenic is heated it will be seen to vaporise without melting, and to sublime along the tube as a lustrous mirror. That portion of the deposit which is nearest to the flame will be seen to be distinctly crystalline, and to have a steel-grey colour, while that which is more remote from the heated part is amorphous in character, almost black in colour, and of a vitreous lustre.

622. To show the combustion of arsenic in oxygen. From

FIG. 191.

3 to 4 grams of arsenic are placed in a piece of combustion tube,

about 30 centimetres long, at a short distance from the end,
and a gentle stream of oxygen passed over it; on heating the
arsenic at a point nearest to the incoming oxygen the element
ignites, and burns with a brilliant bluish-white light. The product of the combustion may be passed through an empty
cylinder, where a large quantity of the arsenious oxide condenses,
and then delivered into a draught flue.

623. The combustion of arsenic in chlorine. This may be
shown in the same manner as described for the combustion of
antimony in chlorine, Experiment 161.

624. The combustion of arsenic in bromine. See Antimony
in Bromine, Experiment 217.

ARSENIURETTED HYDROGEN. AsH_3

625. Preparation by the action of nascent hydrogen upon
arsenious oxide. Hydrogen is generated in a small Woulf's
bottle from granulated zinc and dilute sulphuric acid. The
exit tube should be about 7 millimetres in the bore, and the end
within the bottle should be cut diagonally, in order to prevent
any liquid collecting at the end, and so preventing the flame
from burning steadily. Attached to the exit tube is a piece of

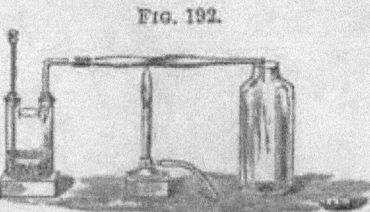

FIG. 192.

hard glass drawn out in the
manner shown in the figure.
When the air is expelled,
the hydrogen is ignited at
the end of the tube, and a
few drops of a solution of
arsenious oxide in hydro-
chloric acid are introduced
by means of the thistle funnel. As soon as the acid solution
of arsenic enters the vessel the rapidity of the evolution of
hydrogen is greatly increased, so that care must be taken that
the gas is only being quite slowly evolved at first, or the reaction

may become rather tumultuous, and the liquid in the bottle
froth over. As soon as the arseniuretted hydrogen finds its
way into the flame, the latter at once assumes a characteristic pale lilac colour. The apparatus should be so arranged
that the products of combustion can be led into a draught
flue.

626. If a cold porcelain dish be depressed upon the flame a
deposition of arsenic at once takes place, as a brownish-black
shining spot.

627. To show the action of heat upon arseniuretted hydrogen.
The hard glass tube is strongly heated at a point where it is
constricted. The arseniuretted hydrogen is decomposed into its
elements, the arsenic being deposited as a mirror upon the sides
of the tube. If the current of gas be not too rapid the whole of
the arsenic will be so deposited, and the flame will no longer
produce a stain upon porcelain.

It will be seen that the arsenic deposit in the tube is
entirely formed upon one side of the flame—namely, on that
side farthest from the generating apparatus.

ANTIMONY

628. The amorphous variety of antimony is prepared by the
electrolysis of a solution of tartar emetic in antimonious chloride.

250 grams of tartar emetic are dissolved in one kilo of
crude antimonious chloride, and the solution placed in a beaker,
which it about two-thirds fills.

The positive electrode should be as large a lump of metallic
antimony as will conveniently hang in the beaker and still
leave room for the negative pole—a piece of the metal weighing
about two kilos will be convenient; a copper wire is soldered

to one end of the lump, which must be suspended in the liquid
to such a depth that the solder is not immersed. By using
so large a positive electrode the solution does not become
weakened, and the deposition of the antimony is therefore
much more rapid. The negative electrode upon which the
antimony is to be deposited may be a stout and clean piece of
copper wire.

The current for the electrolysis is best derived from two
Smee's elements, which should be suspended in large beakers
holding about one or one and a half litres of dilute acid, so that
the experiment may be allowed to go on uninterruptedly for
several days. So arranged, in four days the rod of amorphous
antimony will be about the thickness of a finger; it may then
be removed from the beaker, rinsed in distilled water, and either
hung up to dry, or carefully laid upon blotting-paper.

When struck or scratched, the amorphous antimony undergoes a molecular change, which spreads quickly throughout the
entire length of the rod. This change is attended by a considerable evolution of heat, the temperature rising to 250°C., and
dense fumes of antimonious chloride are evolved.

If a piece of gun-cotton be tied round one end of the bar,
before it is struck at the other end with the edge of a spatula,
the gun-cotton will be inflamed by the heat generated by the
change.

629. To show the combustion of antimony in oxygen. A
fragment of antimony is placed upon a piece of charcoal, having a
shallow hollow scooped out of it, and a spirit-lamp flame directed
upon it by means of a jet of oxygen. The metal almost immediately melts and begins to burn; the lamp is then removed,
and the metal will continue to burn brilliantly in the stream
of oxygen gas.

If while in a state of combustion the globule be thrown upon
a large sheet of cartridge paper, the edges of which have been
folded up in the form of a shallow tray, the globule breaks up
into a number of smaller ones, which run about over the paper,

leaving fanciful streaks, some dotted, some curiously curved, as seen in fig. 193.

FIG. 193.

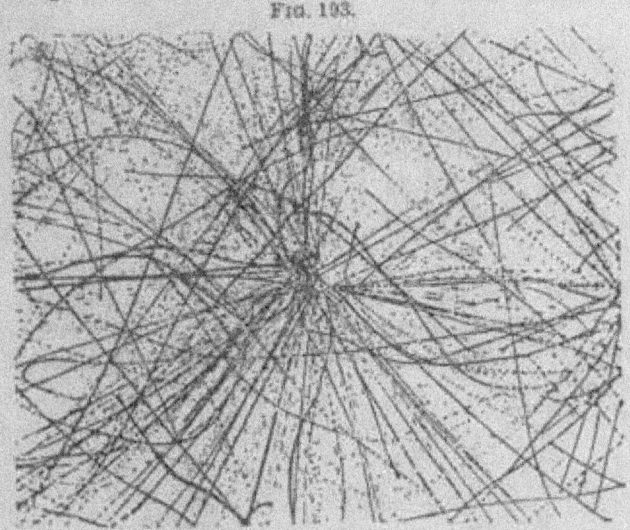

630. The combustion of antimony in chlorine. (See Chlorine, Experiment 161.)

ANTIMONIURETTED HYDROGEN. SbH₃

631. Preparation by the action of nascent hydrogen upon antimonious chloride. This is done in an apparatus similar to that described for the preparation of arseniuretted hydrogen, using a solution of antimonious chloride. The gas burns with a flame resembling in colour that of the corresponding arsenic compound, and when a cold porcelain plate is depressed upon it a similar stain is produced. The antimony stain, however, is blacker than that produced by arseniuretted hydrogen.

632. The decomposition of antimoniuretted hydrogen by heat may be shown in a precisely similar way to that described for arsenic. In this case it will be noticed that the mirror is

formed in the tube much nearer to the flame than was the
arsenic deposit, and also that it is deposited upon both sides of
the heated spot.

633. To show the action of antimoniuretted hydrogen upon
sulphur, in the light. A short glass tube is filled with fragments
of sulphur, about the size of grains of wheat, and supported in
a beam of light from the electric lamp. On allowing antimoniuretted hydrogen to pass through the tube the sulphur
gradually becomes red, owing to the formation of antimony
sulphide.

634. To distinguish between the deposits of arsenic and
antimony they may be treated with a clear solution of bleaching
powder; the arsenic stain is dissolved, while the antimony remains unchanged.

DISSOCIATION

635. Dissociation of iodine.[1]   A small quantity of iodine is
introduced into a test tube, which is then drawn out and
hermetically sealed. The tube is supported in front of a slit
through which a beam of light is issuing, and an image of the
slit projected upon the screen. On gently warming the tube by
means of a Bunsen lamp the violet colour of the iodine vapour
will be seen, but as the temperature is gradually raised the
violet colour gives place to an intense and pure blue. As the
tube cools again the violet tint once more makes its appearance.
Care must be taken to avoid the use of too much iodine, or the
blue colour will be too intense to allow of the passage of the
light; it will be found that ·25 gram for a test tube of about
70 cubic centimetres capacity (i.e. a tube 15 centimetres by
$2\frac{1}{4}$ centimetres) will give a good result. The tube should be
suspended by a wire, so that the whole of it may be heated by
brushing the flame over it.

[1] See Table XXV. in the Appendix.

636. Dissociation of nitrogen peroxide.[1] A quantity of nitrogen peroxide is introduced into a flask, the neck of which has been drawn out, by delivering into it a small quantity of nitric oxide, and the flask is then sealed off. If the flask be cooled and placed in front of the lamp, a pale-yellow disc of light will be seen upon the screen. On heating the flask the colour of the light will rapidly deepen until it has assumed such a dark-brown colour that it appears almost opaque. The flask may be conveniently cooled while in position by allowing a little ether to drop upon a piece of rag or blotting-paper placed upon the flask, and at the same time gently blowing the vessel to hasten the evaporation.

This change of colour may be seen almost equally well by supporting the flask over a white surface, and applying the heat.

637. Dissociation of ammonium chloride. A thin glass tube about 25 centimetres long and 20 millimetres bore has a piece of the stem of a clay tobacco pipe passed through it, and kept in position by two loosely fitting corks. A piece of pure crystallised ammonium chloride is placed in the tube near the middle;

FIG. 194.

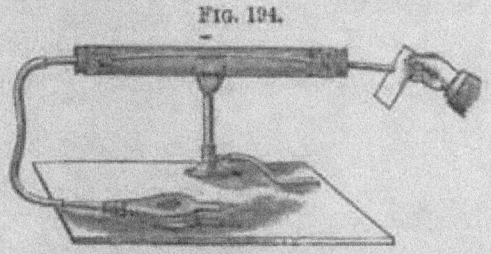

at each end, immediately beyond the cork, is introduced a strip of blue litmus paper, which should be curled round so as to lie flat against the glass tube. One end of the tobacco pipe is connected by means of a piece of caoutchouc tube to a small pair of hand-bellows (fig. 194). Heat is applied to the fragment of

[1] See Table XXVI. in the Appendix.

ammonium chloride, which at once begins to vaporise and to dissociate into the two gases, ammonia and hydrochloric acid, which diffuse in small quantities through the clay pipe stem. Owing to their difference in density, more ammonia passes through the porous stem in a given time than hydrochloric acid, so that the gas within the stem will contain a slight excess of ammonia; if a gentle stream of air be driven through the porous tube by means of the small bellows, and the issuing gas be allowed to impinge upon moistened turmeric paper, the presence of the free ammonia is at once made evident. For the same reason a slight excess of hydrochloric acid gas will be found in the air within the glass tube, and this will make itself evident by reddening the strips of blue litmus paper.

In this experiment it is better to use a pair of bellows, in order to sweep out the gases from the clay pipe, than a stream of hydrogen or other gas, as it is more advantageous to allow the ammonia to accumulate within the pipe and to drive it out at intervals with slight puffs.

638. Dissociation of phosphonium bromide. If the two gases, hydrobromic acid and phosphoretted hydrogen, be passed into a cooled vessel, they will combine and form crystals of phospho-

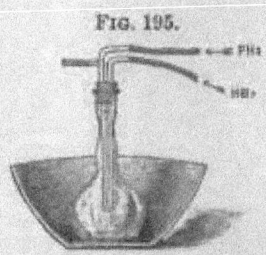

FIG. 195.

nium bromide. The phosphoretted hydrogen may be prepared by the action of phosphorus upon alcoholic potash (see Phosphoretted Hydrogen, Experiment 545), and collected in a gas-holder. The hydrobromic acid is best prepared by passing hydrogen and bromine over a heated platinum spiral (see Hydrobromic Acid, Experiment 225), as the current of gas can be regulated to any desired rate.

The compound is conveniently collected by delivering the two gases into a flask, which should be cooled by being placed in a freezing mixture. If it be desired to preserve a quantity of the compound as a specimen, the neck of the flask may be constricted as seen in the figure, and the tubes delivering the gases drawn out sufficiently fine to pass the constriction of the neck. Immediately the gases are mixed, a deposit of the phosphonium bromide begins to collect upon the sides of the flask, and in a short time a considerable quantity of the compound will be formed; the gases issuing from the exit tube may be allowed to escape into a convenient draught.

For the purposes of the experiment for which the phosphonium bromide is being prepared it will be found convenient to transfer small quantities of it to a number of small U tubes, in which it can be sealed up and preserved until required for use.

For this purpose the flask is removed from the freezing mixture, and a gentle stream of hydrogen is passed in by one of the two long tubes, the other one being closed by placing a pinch-cock upon the caoutchouc tube. One of the U tubes is connected with the exit tube, and is immersed in a freezing mixture. In this way a quantity of the compound will be condensed in the U tube, which may then be sealed up. When several of the tubes have been in this way prepared, the flask itself may be sealed up at the constricted part.

To show the dissociation of the phosphonium bromide, a glass tube 30 centimetres long and 1 centimetre bore has two small branch tubes blown into it, one near to each

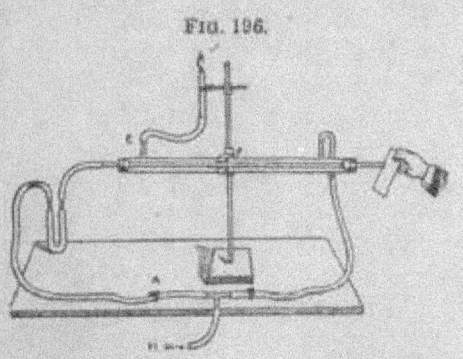

FIG. 196.

end. The long stem of a clay tobacco pipe is inserted into the tube, being fitted at each end by means of short pieces of caoutchouc tube, and being long enough to project one or two centimetres beyond the glass at each end (fig. 196). (As the stem of a

U

pipe is not often straight enough throughout such a length as this to pass down a narrow tube, it may be necessary to join together two shorter pieces by means of a small piece of caoutchouc tube.) A stream of hydrogen is passed through the glass tube, and the gas inflamed at an ordinary fish-tail burner, F, screwed into a short piece of lead pipe and connected to the exit tube, E, of the apparatus, the supply of hydrogen being so regulated that a small flame is burning, and, the burner being of metal, the flame will be non-luminous.

A U tube containing the phosphonium bromide is attached to one end of the tobacco pipe, the other limb of the tube being connected with a hydrogen supply. This may be conveniently arranged as seen in the figure. By opening the screw-cock A, a gentle stream of hydrogen can be caused to pass through the U tube, and almost immediately the flame will show the presence of phosphoretted hydrogen, this gas having diffused through the clay pipe into the glass tube. A piece of litmus paper held at the opposite end of the tobacco pipe is instantly reddened. If the gas issuing from this pipe be inflamed, it of course burns with a luminous flame due to phosphoretted hydrogen; but if the flame at F be blown out, and the issuing gas allowed to impinge upon litmus, no reddening will result; the amount of hydrobromic acid, if any, which will have diffused through being so extremely small.

639. Dissociation of phosphonium chloride. By passing a mixture of phosphoretted hydrogen and hydrochloric acid gas through a U tube contained in a freezing mixture, the phosphonium chloride is deposited as a white crystalline solid, which may be preserved as such by sealing the tube. By slightly cooling one limb of the sealed tube and leaving the other exposed, the substance may be readily sublimed and obtained in the form of large and perfect crystals.

If a mixture of these two gases, in equal volumes, be introduced into a Cailletet tube (see Liquefaction of Gases, Experiment 654) and subjected to compression, crystals of phosphonium chloride are formed upon the sides of the tube; on releasing the pressure the crystals dissociate again into the original gases. If this experiment be performed in front of the lamp, an image of the tube being thrown upon the screen, it is advisable, on account of the heating of the tube by the light, to slightly cool it, either by surrounding it with cold water or by dropping a little ether upon it; for if the temperature be allowed to reach 25°C. the compound is obtained in the liquid state—that is to say, the substance melts at about that temperature—and therefore no formation of crystals takes place, although the pressure be increased to as much as 150 atmospheres. If while the substance is in the liquid condition and under considerable compression the pressure be suddenly released, the reduction of temperature which results causes the instant solidification of the compound, and a deposition of fine snow-like crystals is seen throughout the length of the tube.

640. If a Cailletet apparatus be not available, the formation of the phosphonium chloride by pressure, and its subsequent dissociation, may be shown by means of a column of mercury. A stout glass tube about 70 centimetres long is drawn out at one end, and connected with a mercury reservoir by means of a long piece of webbed caoutchouc tube. The other end is drawn out to a fine point. Mercury is placed in the reservoir, which is then raised until the metal begins to flow out from the point of the glass tube. The tube is then connected by means of a piece of caoutchouc to a small vessel containing the mixed gases, over mercury. A quantity of the gas is drawn over into the tube by lowering the reservoir, and the tube is then sealed off. The upper part of the tube is surrounded by a wider tube, as shown in the figure, in order to cool the contained gas. If the tube be cooled to 0°C., by means of ice-cold water, the pressure

FIG. 197.

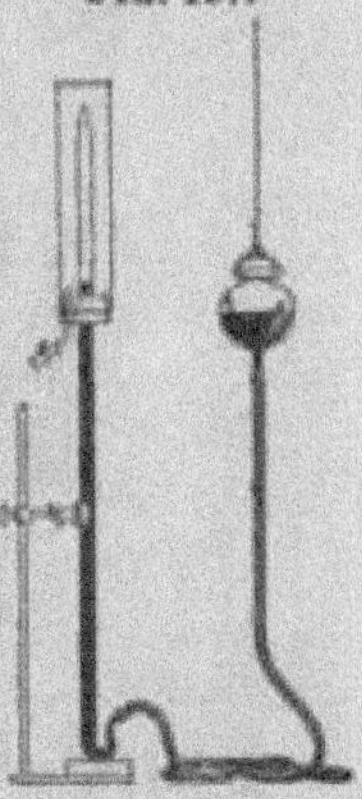

required to effect the crystallisation of the compound will be 13 atmospheres, the reservoir in this case having to be raised to a height of 10 metres. By cooling the tube below this temperature, which may be conveniently done by surrounding it with brine previously cooled in a freezing mixture of ice and salt, the crystallisation may be brought about at much lower pressures, as the following figures will show :—

At − 5°C., mercury column required = 7 metres, being a pressure of about 9 atmospheres.

At − 10°C., mercury column required = 5 metres, being a pressure of about 6·5 atmospheres.

At − 15°C., mercury column required = 3 metres, being a pressure of about 4 atmospheres.

As a temperature of − 15°C. is readily obtained in this way, it will be found most convenient to cool the tube to this degree, and thus avoid the use of any excessive length of flexible tube.

641. Dissociation of phosphorus pentachloride. A few grams of phosphorus pentachloride are sealed up in a wide glass tube. On applying a moderate heat to the compound it undergoes dissociation, and the yellow colour of the chlorine will be apparent. As the tube cools again the contents become colourless as before.

642. Dissociation of ammonium oxalate. A quantity of ammonium oxalate is dissolved in water, and a little litmus added. If the salt is pure the solution will be neutral and the litmus will retain its violet colour, otherwise the solution must be rendered perfectly neutral.

The liquid is then boiled in a small retort, the stem of which is made to dip into water which has been coloured with reddened litmus.

In the course of a few minutes the reddened litmus in the receiver is turned blue, while the solution in the retort becomes acid, which is seen by the litmus becoming red.

LIQUEFACTION OF GASES

643. To show the effect of pressure and of cold successively upon the same sample of gas. A suitable gas to employ is ether vapour. A glass tube about a metre long is connected by means of flexible tube to a mercury reservoir (conveniently made from a dropping funnel), in the manner shown in fig. 198. The upper end of the tube is drawn out to a fine capillary tube, which is bent over so as to dip into a small vessel containing ether. The reservoir is raised until the air in the apparatus is expelled, and then on slightly lowering it, about 2 c.c. are drawn over into the tube. The tube is then sealed off by directing a fine blowpipe flame upon it at a distance from the end. The tube is then jacketed with a wider tube, the cork at the bottom also carrying a small exit tube which can be closed by means of a pinch-cock and piece of rubber tube. Warm water is poured into the outer tube so as to cause the ether to completely vaporise while the reservoir is at a point a little below the top of the tube. On raising the reservoir, and so subjecting the vapour to increased pressure, the ether will be liquefied; on lowering it to its former position the ether again returns to the gaseous state. The warm water is now withdrawn by opening the pinch-cock (the mercury reservoir being allowed to remain stationary) and replaced by cold water, preferably iced. As the cold water is applied the mercury once more rises in the tube, and the ether vapour is again condensed to the liquid condition.

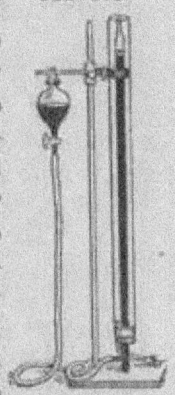

FIG 198.

The most serviceable flexible tube for conveying mercury and withstanding pressure consists of caoutchouc tubing, 2 millimetres bore and 2 millimetres thick in the walls, upon which has been woven a covering of cotton. Such tube is more durable than uncovered thick wall tube, and will stand a much greater internal pressure.

644 To show the liquefaction of a gas in a sealed tube (Faraday's method).   An apparatus slightly modified from the usually described form of Faraday tube is made by sealing a short piece of narrow tube upon one end of a tube about 23 centimetres long, and of about the substance of ordinary 'combustion' tubing, the narrow tube being then bent at a right angle, as shown in the figure.   A quantity of dry silver chloride is placed in the wide part of the tube while the end is still open, a little plug of asbestos being first pushed up to the joint in order to prevent any fragments of the chloride from passing into the narrow tube.   When nearly full the tube is drawn off, and a stream of dry ammonia gas passed through for twenty or thirty minutes.   The tube is then sealed up at both ends.   The narrow limb is placed in a freezing mixture, and a gentle heat applied to the other portion by brushing a Bunsen flame backwards and forwards along it.   The ammoniacal silver chloride melts and gives up its ammonia, which becomes condensed by its own pressure and the cold of the freezing mixture, and collects as a colourless liquid in the narrow limb.   On removing the tube from the freezing mixture and allowing it to stand, the ammonia boils off into vapour, and is reabsorbed by the silver chloride. The tube should be held in a horizontal position, so that the fused contents may present as much surface as possible to the ammonia.

FIG. 199.

645. The power of charcoal to absorb gases may be made use of for obtaining liquefied ammonia.   In this case the tube is packed with charcoal in small grains, and a stream of dry ammonia gas passed over it.   The tube should be first heated, and then allowed to become quite cold while the gas is still passing through; it is then sealed up and treated exactly as described above.

It is convenient, even when silver chloride is used to absorb the ammonia, to mix it with sufficient charcoal to prevent the mass from running together when heated.

646. Liquid sulphur dioxide may also be produced by absorbing the gas by means of charcoal and heating the material in a sealed tube.

647. The liquefaction of an easily condensed gas may be effected by the pressure obtained by a column of mercury. A convenient gas for the purpose is sulphur dioxide. A U tube, about a metre long, having a small branch tube blown into the bend, is connected by this branch to a mercury reservoir by means of a long flexible tube. The ends of the U tube are drawn out so that gas may be introduced, and so that they may ultimately be sealed off. To fill the tubes, mercury is poured into the reservoir, which is then lowered down so that a small quantity of the metal remains in the bend of the U tube. A stream of sulphur dioxide is then driven down one limb, so that it will bubble past the mercury in the bend and escape at the open end of the other limb. When all the air is thus swept out, the tube into which the gas is being delivered is sealed by a blowpipe; the reservoir is then raised until all the gas in the open limb is driven out, and on again lowering the mercury air will re-enter this tube. After adjusting the level of the mercury in the two limbs, the second tube, now containing air, is sealed up. This tube is to serve as an indicator of

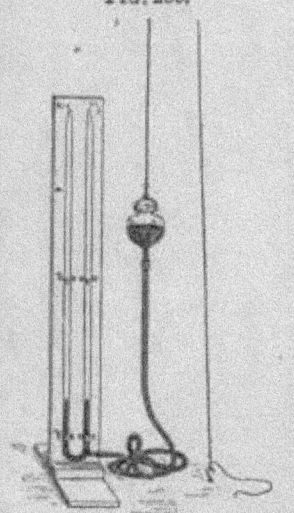

Fig. 200.

the pressure, and is roughly graduated by means of india-rubber rings. On raising the reservoir, by means of a cord passing through a pulley fixed at a convenient height, the mercury will be seen to rise at first equally in both limbs; but as the pressure reaches 3 to 4 atmospheres the sulphur dioxide is condensed to a small quantity of liquid, which will occupy the narrow part of the tube at the top. On again lowering the reservoir this liquid boils off once more into gas, and occupies the original volume in the tube.

648. The liquefaction of a gas requiring only a few atmospheres pressure to condense it may be effected by forcing the mercury into the tube by means of a plunger. In the case of such an easily condensible gas as sulphur dioxide the entire apparatus may be constructed of glass. Fig. 201 represents a

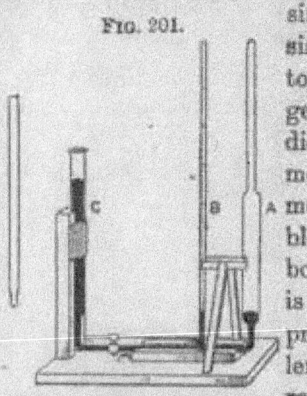

FIG. 201.

simple form of the apparatus; it consists of three vertical tubes connected together by a horizontal piece of Sprengel tube. Tube A contains the sulphur dioxide; this tube is constructed of a moderately stout tube about 15 centimetres long, upon one end of which is blown a piece of tube having a flat bore. The object of this flat bore tube is two-fold. The amount of liquid produced will occupy a much greater length in the tube than if the bore were round, and so allow the experiment to be made on a much smaller scale; and also it renders it possible to project an image of the tube, with the little column of liquid gas, upon the screen. Tube B is a pressure gauge, and consists merely of a piece of Sprengel tube graduated by means of caoutchouc rings. Tube C is the reservoir for the mercury, and must be of such a size that it can contain enough mercury to fill the rest of the apparatus. A glass stopcock is blown into the horizontal tube, so that communication between the reservoir and the other tubes may be closed or opened at will. The whole apparatus may be conveniently mounted on a wooden stand, as seen in the figure. To fill the apparatus a quantity of mercury is poured into tube C, the cock being closed. A small quantity

of mercury is then allowed to pass the cock, so as to fill the
horizontal tube and rise a very short distance in the two tubes
A and B.   A stream of sulphur dioxide is next passed in
through the drawn-out and open end of tube B, the gas bubbling through the mercury and escaping at the drawn-out open
end of A.   When the air is entirely displaced the point of
tube A is sealed up.   The sulphur dioxide in the tube B may
be allowed to diffuse out through the open end, or may be
swept out by driving the mercury up the tube by means of the
piston or plunger, and when the tube is filled with air it is also
sealed up.   The piston may be conveniently made of a very
stout glass rod which has been constricted near one end.   A
short piece of caoutchouc tube is slipped over this rod at the
constriction, and by adjusting its position upon the slightly
taper part of the rod the piston may be made to fit the tube C.
The piston is moistened with a little glycerine, which gives it a
perfectly easy movement against the sides of the tube.   As the
piston is pressed down into the tube, the cock being open, the
sulphur dioxide in tube A is easily liquefied, the air-gauge
registering the pressure, and the liquid can be maintained in
the tube by closing the cock.   By gently opening the cock and
again releasing the pressure, the sulphur dioxide once more
boils off into gas.   When once charged, the apparatus may be
preserved for use for an indefinite time, the piston being withdrawn, and the tube C being closed by a cork.

649. To liquefy a gas by the application of cold only.   The
most convenient form of apparatus for this purpose is the condensing tube shown in fig. 202.   It consists of an elongated bulb,
blown upon the end of a glass tube of about 1 centimetre
bore and 20 centimetres long; into the shoulder of the bulb
is blown a piece of small tube, which is brought up parallel
with the wider tube and bent as seen in the figure.   The tube
is immersed in a suitable freezing mixture, and the gas passed
into it by the small tube, in order to expose it to as much cooling surface as possible.   The advantages of this form of con-

densing tube over the ordinary apparatus are, first, that a much smaller quantity of freezing mixture is necessary; and, second,

FIG. 202.

that the gas is more rapidly condensed, owing to the small tube through which it enters. When the freezing mixture to be used is ice and salt, or ice and chloride of calcium, the mixture may be contained in a glass cylinder; but when solid carbon dioxide and ether is used, a convenient receptacle for it is a boiling tube of such a size that the condensing tube can just pass easily down. The boiling tube should be wrapped in either flannel or green baize. A quantity of ether, about half filling the tube, is introduced, and fragments of solid carbon dioxide added. In this way, with the minimum expenditure of carbon dioxide, considerable quantities of such gases as ammonia, chlorine, &c., may be readily liquefied.

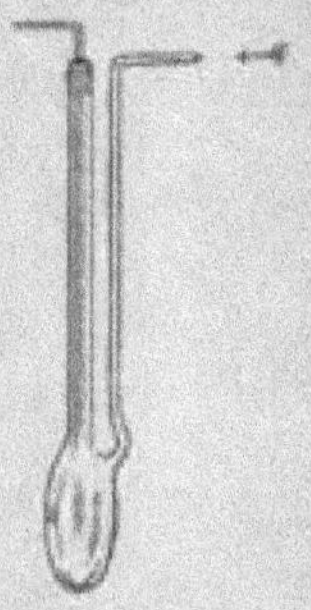

650. To show that a gas suffers a change in temperature by either sudden compression or expansion. In the first case, the heat generated by compression may be demonstrated by igniting a mixture of air and carbon disulphide vapour by suddenly sub-

FIG. 203.

jecting it to pressure. This is a modification of the familiar tinder experiment. A glass cylinder, made of a piece of glass tube with very thick walls, is fitted with a steel plunger, the latter being provided with a leather washer, as in the case of an ordinary pump. A minute fragment of blotting-paper, mois-

tened with carbon disulphide, is dropped into the cylinder, and
the plunger very suddenly thrust down; a feeble flash of light,
which will be distinctly visible in a darkened room, will be seen
within the cylinder.[1]

651. The opposite effect—viz. the production of cold by the
sudden expansion of a gas—may be shown by placing the bulb of

[1] It is a remarkable and interesting fact that the principle involved in this
experiment is actually made use of by certain savage tribes of North Borneo
as a means of obtaining fire. The figure represents a specimen of the complete apparatus collected by Mr. Skertchly, and presented by him to the
Anthropological Museum of Oxford.

The cylinder A is made of an alloy of 2 parts of lead to 1 part of tin, and
is cast in a bamboo mould, the mould being a thin piece of bamboo split
lengthwise, and on the interior of which the ornamental bands, &c., are in-
cised. The cylinder is usually made 8½ centimetres long, 12 millimetres
diameter, and 9 millimetres bore.

FIG. 204.

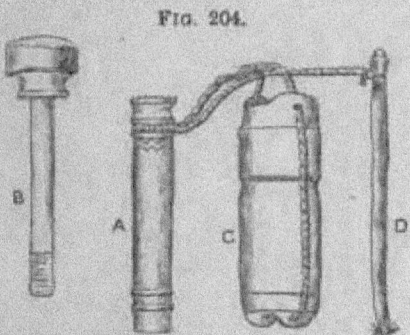

The piston B is made of any hard wood, has a head or knob at the top, and
is packed at the end, for 2½ centimetres, with cloth, to render the apparatus
air-tight. The end is slightly hollowed for the reception of the tinder.

The tinder which is found to answer best is made from the external
covering of the stem of a particular palm; it is carried in the tinder-box C,
which is constructed out of a joint of a bamboo.

D is the cleaning stick, and this and the tinder-box are attached to the
cylinder by means of a piece of string.

A small piece of tinder is placed in the hollowed end of the piston, which is
then inserted in the mouth of the cylinder. Holding the cylinder in the left
hand, the knob of the piston is smartly struck with the open right hand with
sufficient force to drive the piston home. The piston is instantly and quickly
withdrawn, and the tinder is seen to be alight.

an air-thermometer beneath the receiver of an air-pump, and suddenly rarefying the air by one or two strokes of the pump. For this purpose a small tubulated bell-jar has fitted into it a glass bulb about 20 mm. in diameter; the stem of the bulb passes through a caoutchouc stopper which tightly fits the neck of the bell-jar. It is well to smoke the bulb after it is fixed into its place, which may be done with the flame of a burning taper. The bell is then placed upon the plate of an air-pump, and the stem of the bulb is connected by a piece of fine caoutchouc tube to a straight piece of flat bore tube (shown in section in the figure), which dips into a small vessel contain-

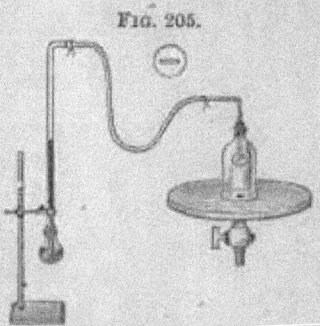

Fig. 205.

ing some coloured water. The flat bore tube is placed in front of a lantern and an image of it thrown upon the screen, the coloured liquid being brought up in the tube to any convenient point by simply squeezing the caoutchouc connect-ing tube for a moment, thereby expelling one or two bubbles of air. When the liquid is just within the field, one or two strokes of the air-pump are made. The sudden rarefaction of the atmosphere within the bell-jar has the effect of lowering the temperature of the remaining air, and consequently cooling the blackened bulb; the effect of the cooling of the bulb is to contract the air within it, and cause the liquid in the stem to move in a direction towards the bulb, i.e. to rise in the stem. When the liquid has risen to its maximum height, and the movement stopped, if air be allowed rapidly to re-enter the bell-jar the reverse action will be seen to take place, the bulb being warmed by the rarefied air suddenly becoming more dense. It will be found convenient to attach permanently to the flat bore tube a small vessel to contain the coloured liquid; for this purpose a minute flask, or small bulb, may be fitted upon the end of the tube by means of a small piece of caoutchouc tube; the neck of the bulb being of such a size that the tube with the rubber upon it can be squeezed into it; a small hole is then punctured in the bulb. (See also the Atmosphere, Experiment 341.)

652. The reduction of temperature consequent upon the expansion of a gas when released from high pressure may be shown by allowing the gas to escape from a cylinder of compressed oxygen, and causing it to blow against the bulb of an air thermometer. For this purpose a short piece of small brass pipe is soldered into the nozzle of the gas-bottle. The projecting end of this pipe is then closed by means of a little solder, and a fine pin-hole bored through the end with a small drill. The gas-bottle is then supported in a horizontal position, and the bulb of an air thermometer (see fig. 50) is held in a clamp quite near to the jet. On opening the valve, the outrushing gas will have its temperature lowered by its sudden expansion, which will be immediately evident by a disturbance of the level of the liquid in the bent limb of the thermometer. This effect can also be obtained, although in a less pronounced degree, without constructing a special jet, but by removing the nozzle altogether from the bottle and bringing the bulb as near to the valve as possible before releasing the gas.

653. The liquefaction of gases by means of Oersted's condenser This apparatus consists of three parts; viz., first, a cylinder of very stout glass fixed into a wooden base, and with a brass flange fitted upon the top; second, a small brass pump fitted with a three-way cock, and a small glass vessel capable of holding about 200 c.c. of water. This little pump is screwed into a brass base, which constitutes the cover for the cylinder, and which accurately fits the flange, being so made that on turning it a few degrees these two portions of the instrument are secured together. The third part is a small iron pot attached to the end of a brass rod, which also carries a disc of metal perforated with two or three holes. This pot is filled with mercury, and the tubes containing the gases to be liquefied are

stood in it, having been first passed up through the holes in the disc which keep them in position.

The tubes may be made of thin glass, drawn off to a long

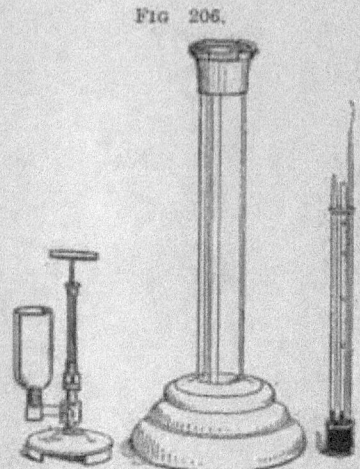

FIG. 206.

taper end in order to render the condensed gas more visible. The tubes must be of such a bore that the little reservoir in which they stand will contain enough mercury to more than fill all that are to be used at once.

A piece of Sprengel tube closed at one end, and containing air, should be placed in the apparatus to serve as a gauge to indicate the pressure; this tube may be graduated by small india-rubber rings. To keep the tubes steady, and with their ends depressed to the bottom of the mercury in the small reservoir, small pieces of cork may be packed into the holes in the disc through which they pass, or a ring of india-rubber may be slipped over them and brought down close to the disc. When the tubes are in place, the whole of this part of the apparatus is lowered down into the glass cylinder, which has been previously nearly filled with water. The little condensing pump is then fixed into position. With each stroke of the pump, the three-way cock has to be made use of, for when raising the piston the small side reservoir is placed in communication with the pump, and when pushing it down communication with the reservoir is cut off, and that with the main cylinder opened.

In this way a pressure of ten or twelve atmospheres may readily be obtained, without any danger to the cylinder. If it be required to project an image of the tube or tubes upon the screen, the glass cylinder must be encased in a metal jacket having parallel sides of plate glass. The jacket, or case, is constructed in two halves, which are bolted together by a number of screws, and rendered water-tight by means of india-rubber packing. This case is filled up with water, when it will be possible to obtain an image of the tubes within the cylinder.

654. The liquefaction of gases by means of Cailletet's apparatus. This apparatus consists of a small hydraulic pump (fig. 207), and a strong steel bottle (fig. 208), which can be put into communication with each other by means of a fine copper tube. The gas to be experimented upon is contained in a glass tube,

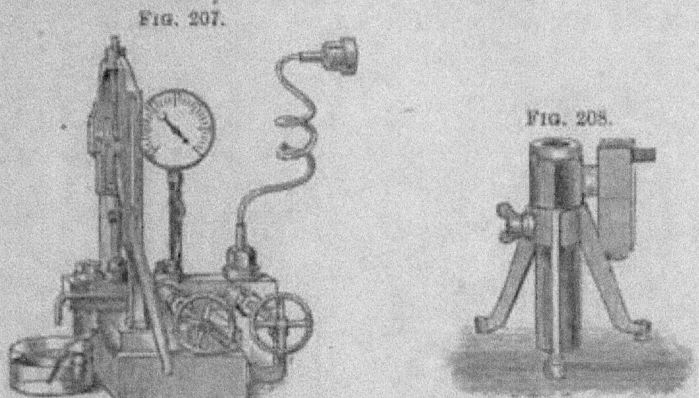

FIG. 207.

FIG. 208.

having a gun-metal collar cemented upon it, and which is lowered into the steel bottle, into which a quantity of mercury has been poured. On working the pump, water is forced into the steel bottle, and thereby drives the mercury up into the glass tube, thus compressing the gas. For lecture purposes this is a very useful piece of apparatus, having rendered the Oersted's condenser practically obsolete, and as it requires some little care for the successful manipulation of it, its use may be described at some length.

To cement the glass tubes into the gun-metal collars. This may be done with marine glue, or better with bicycle cement, which for many laboratory purposes is an extremely useful cement. The metal piece is heated by being placed in a pan of boiling water, and that part of the glass tube to which the collar is to be fastened is smeared with a quantity of the cement by gently warming over a flame. The hot piece of metal is then removed from the water and placed upon a small retort-stand ring, or held in a clamp, while the tube is slowly pushed through. It is then left in this vertical position until quite cold, when any excess of cement can be removed with a warm knife, and the parts finally cleaned by a little benzene, or carbon disulphide.

To fill the tubes with gas. The closed end of the tube is opened by cutting off the extreme point with a file, and three or four cubic centimetres of mercury are put into the tube by dipping the recurved end into a vessel containing mercury; the tube is then held in a clamp in a nearly horizontal position, and a stream of the gas with which the tube is to be filled is passed through by means of a piece of caoutchouc tube. When the air is all displaced the point is sealed by a fine blowpipe flame, and the tube raised to a vertical position. The mercury which now fills the recurved end forms a trap to prevent the ingress of air, while removing the tube to a small vessel of mercury in which it is placed until required. It is convenient to have a rack capable of holding a number of tubes which have been filled with different gases; such a contrivance is shown in fig. 209, where each tube is stood in a small beaker containing mercury.

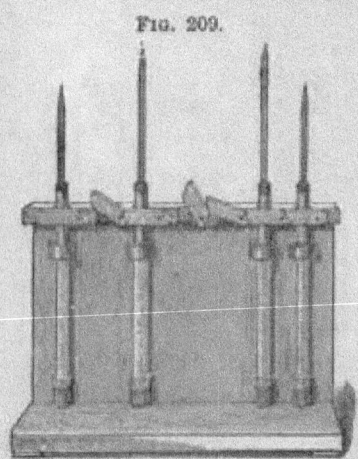

Fig. 209.

The quantity of mercury which must be put into the steel bottle must be ascertained once for all by trial. If too little be present, water will get into the gas tube, which in many cases by its action upon the gas would be fatal to the experiment. If too much mercury be used, it is liable to find its way along the

copper tube into the pump, which will rapidly become corroded
by it.   Before the steel bottle is connected to the pump,
such a quantity of mercury is introduced that when a tube
is lowered into its place the excess will overflow from the
side aperture, the bottle being slightly tipped towards that side;
two or three different tubes should in this way be lowered into
the bottle, as they are liable to vary slightly in size.   When
nothing more is forced out from the opening, the tube is again
removed, and the mercury which is in the bottle is poured out
and measured.   It is a good plan to adopt the additional precaution of having a mercury trap between the bottle and the
pump.   This consists of a small steel cylinder which is screwed
directly to the bottle, coming between it and the copper pipe.
This cylinder will receive any mercury which might overflow
from the bottle, and being provided with a screw plug in the
bottom, it may be opened from time to time (fig. 208).

When the apparatus is to be used, the tube containing the
gas is lowered into its place (the collar sitting upon a leather
washer placed upon the shoulder within the
bottle) and is securely held down by a nut,
which passes over the narrow part of the
collar, and engages in the thread upon the
neck of the bottle.   If the gas to be liquefied
does not require more than about 150 atmospheres pressure, the tube need not be pro-
tected by any outer jacket, as all such guard
tubes render it more difficult to see the experiment, and if an image of the tube is being
projected upon the screen they greatly dis-

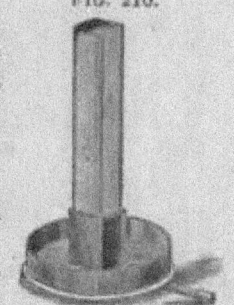

Fig. 210.

tort the image.   When it is needful to surround the tube,
either for safety or in order to cool or warm the gas, the jacket
should be a tube made by cementing flat glass together in the
form of a square tube, or long square cell, fig. 210; in this way
the image will be the least obscured.   As the pump is put into
action the mercury will be forced up into the narrow part of the
glass tube, and the liquid will make its appearance as a small

x

column floating upon the mercury. If the pressure required to liquefy the particular gas under consideration be not greater than about 200 atmospheres, the operation may be performed entirely by the use of the long lever handle of the pump; the side piston, which is actuated by the small wheel, being only intended to enable the experimenter to exert additional pressure upon the gas when the use of the lever becomes uncomfortably hard work. To release the pressure, water is allowed to escape by turning the second wheel. If this be done gently the liquefied gas will be seen to boil off slowly into gas; if done suddenly the liquid will instantly flash off into gas, thereby developing a degree of cold which in many cases causes the formation of a cloud of either the liquefied or solidified gas. With this apparatus, and without the application of any refrigerating mixtures, a number of gases may be reduced to the liquid state, gases whose critical points are not below the ordinary temperature, and which require not more than about 150 atmospheres for their liquefaction.[1]

655. When the critical point[2] of a gas is only a few degrees below the ordinary temperature, it is possible to obtain it in the liquid state without the use of any refrigerators, by cooling it by sudden expansion. Ethylene is a good illustration to employ. When this gas is compressed even to 150 atmospheres no liquefaction takes place, but if the pressure be slightly but suddenly released, by a rapid movement of the wheel, the internal temperature will be lowered below the critical point (about $+10°C.$), and the gas will be liquefied.

By a little manipulation with the pump, even a gas whose critical point is as high as that of carbon dioxide ($31·9°C.$) may be made to show a similar result; by very rapidly compressing this gas its temperature may be raised above this point, and no liquid will be produced even at 150 atmospheres, but on a momentary release of the pressure the temperature falls and the gas is condensed.

[1] See Table XXXII. in the Appendix.
[2] See Table XXXI. in the Appendix.

The critical point of a gas, *e.g.* carbon dioxide, may also be shown by surrounding the tube with the square glass cell, into which warm water is poured having a temperature a few degrees above the critical point. On compressing the gas with a pressure greatly above what would be required at ordinary temperatures, it will be seen that there is no liquefaction taking place. If the warm water be run out and replaced by cold water, the liquid state is at once assumed by the gas.

656. Liquefaction by 'self-cooling.' The most recent methods for the liquefaction of gases whose critical temperatures are very low, such as air or oxygen, are based upon the principle illustrated in Experiments 651 and 652. The gas under a high pressure (in the case of air or oxygen from 140 to 200 atmospheres) is made to issue from a fine orifice at one end of a copper spiral tube which is insulated from the surrounding air as perfectly as possible. The gas, cooled by its own sudden expansion, is caused to sweep over the spiral, whereby the issuing gas is itself cooled before reaching the orifice. Each succeeding portion of gas therefore undergoes its expansion from a lower platform of temperature than that which escaped immediately before it, and consequently the temperature of the gas gradually and steadily falls until at length the point is reached at which liquefaction begins—*i.e.* the boiling-point of the liquid is reached. To keep up a supply of gas under such a high pressure necessitates the employment of a powerful compressor, driven by a suitable engine.

If arrangement is made to first cool the apparatus by means of solid carbon dioxide—as in Dewar's apparatus—a moderate quantity of liquid oxygen can be obtained by substituting for the compressor and engine, a large steel cylinder of oxygen compressed to about 160 atmospheres.

In either case it will be evident that the liquefaction of air or oxygen by this 'self-cooling' method either involves machinery and apparatus which is beyond the resources of all except the very best equipped laboratories, or necessitates the laboratory being within convenient reach of the manufacturer of compressed gases.

The principle of these *liquefiers*, and the actual liquefaction of a gas by cooling due to its own expansion, may however be illustrated by means of the following simple piece of apparatus,

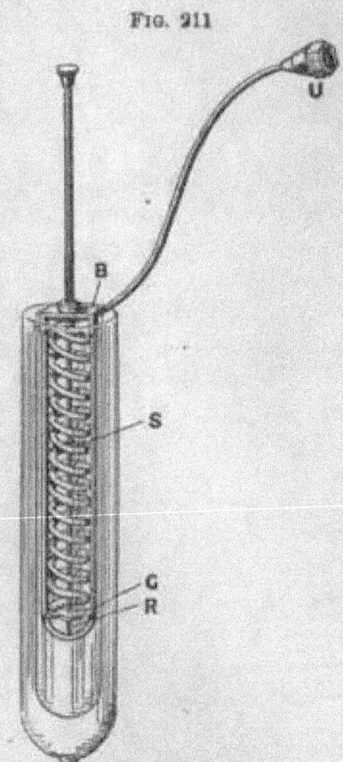

Fig. 211

the gas to be employed for the purpose being nitrous oxide. The apparatus consists of a simple spiral tube s, fig. 211, made of fine drawn copper tube. It is necessary that *drawn* tube should be employed, and not tubing which has a brazed seam, as in the latter case the seam is very liable to open when the tube is bent into a spiral. The bore of the tube should be $1\frac{1}{2}$ millimetres in diameter, and the walls about 1 millimetre thick.

The lower end of the tube is closed, and is bent up as shown in the figure. The orifice from which the gas is to escape is a small hole bored in this bend, which can be opened or closed at will by means of the fine-pointed steel rod R, which has a thread cut upon the thicker part near the top whereby it can be screwed up and down. In order to keep it central, it passes through a hole in the guide-bar G.

Rigidity is given to the spiral by the two narrow strips or bands of brass soldered down opposite sides. These are bridged across the top by the bar B, through which the steel rod passes.

The apparatus is attached to the nitrous oxide cylinder by a union U. The entire apparatus can readily be made by any intelligent laboratory mechanic or fitter.

The rest of the apparatus consists of a vacuum-jacketted glass tube; i.e. a double-walled glass tube in which the space between the walls has been rendered perfectly vacuous.

This tube should be of such a size that the spiral can just pass down, and reach to within about 30 centimetres of the bottom.

To perform the experiment, the copper tube is attached to a steel cylinder of nitrous oxide, the latter being supported in a vertical position; and the glass tube, held in a clamp, is brought up underneath the spiral. The valve of the little apparatus is screwed down so as to entirely close the orifice, and the valve of the gas cylinder is fully opened. If all is tight, there should be no issue of gas under these conditions.

The orifice of the spiral is now gently opened by slightly unscrewing the valve, when the gas makes its escape from the jet under a pressure of 60 or 70 atmospheres. The gas, cooled by its expansion, and brushing past the spiral as it makes its way out of the glass tube, cools the copper tube, and in two or three minutes the temperature of the latter will become lowered to the boiling-point of the gas when a spray of liquid will be seen issuing from the jet along with the gas. In a very short time several cubic centimetres of liquid will be collected at the bottom of the glass tube. If desired, an image of the jet and lower part of the glass tube can be projected upon the screen.

As supplied to the market, for use by the anæsthetist, nitrous oxide is contained in steel bottles or cylinders which are not fitted with a tube inside reaching down to the bottom as in the case of carbonic acid cylinders.[1] Therefore, although the cylinder contains the gas already in the liquid state, it is not

[1] The method described in Exp. 308 for collecting liquid nitrous oxide requires that the liquid be contained in a carbonic acid cylinder.

the *liquid* which is driven out of the bottle, but merely the *gas* evolved by the evaporation of this liquid on release of the valve. Indeed, should the liquid by any mistake be contained in a cylinder having such a syphon tube, or what comes to the same thing, if the cylinder be supported in an inverted position so that the liquid itself is forced direct into the spiral tube, the experiment will inevitably fail; because the reduction in temperature caused by the evaporation of the first few drops which escape from the jet *immediately freezes* the liquid in the lower turns of the spiral tube and thus effectually stops it up— the freezing point of nitrous oxide being only about twelve degrees below the boiling-point. It is owing to the still closer proximity between the melting and boiling-points in the case of carbon dioxide that this gas cannot be used for this experiment in place of nitrous oxide.

As the liquid contained in the cylinder evaporates to furnish the supply of gas, obviously its own temperature will begin to fall in consequence, and in proportion as this takes place the *pressure under which the gas issues from the jet* will also become less. Now, since the success of the experiment depends upon this pressure being maintained as high as possible, it is necessary to prevent the steel cylinder from getting cool during the operation. If the cylinder be moderately large, the mass of metal will usually itself prevent the undue cooling of the contained liquid; but in the case of a small steel bottle it must be stood in a jar or other convenient vessel containing lukewarm water— water of a temperature between 20° and 25° will be amply warm enough.

The solidification of nitrous oxide by its own evaporation may readily be shown by directing a few puffs of wind upon the surface of the liquid in the glass tube by means of a blow-pipe bellows. The liquid immediately freezes on its surface, allowing of the tube being inverted without the escape of any of the liquid.

EXPERIMENTS ON ELECTROLYSIS.

657. A number of experiments involving the electrolytic decomposition of compounds are best shown by projection upon the screen, the electrolysis being conducted in a glass cell with parallel sides. Such a cell is shown in fig. 14, page 18.

A simple modification of this cell may readily be constructed by squeezing a piece of caoutchouc tube between two pieces of sheet glass; the whole being secured together, without the aid of any cement, either by binding-wire (fig. 212) or by means of

FIG. 212.

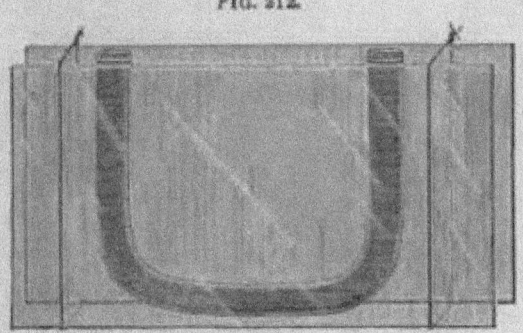

two strips of band-brass which can be clamped together by means of small bolts and nuts passing through their ends. A foot for the cell is made by nailing two strips of wood upon a small piece of board so as to form a groove into which the cell will just fit (fig. 213).

FIG. 213

The electrodes may be conveniently fitted to such a cell by sealing the wires into the ends of short glass tubes of small bore, which are passed through corks cut slightly wedge-shape, so that they can easily be fitted into the cell. The glass tubes are nearly filled with mercury, and connection with the battery is made by dipping the wire leads into the mercury (fig. 214).

Fig. 214.

If the cell is a very small one, intended to be used with lenses of high magnifying power, as in the lantern microscope (see p. 327), the electrodes must be rather close together. They should therefore be both passed through the same cork.

658. Electrolysis of lead acetate solution. A strong solution of lead acetate is placed in such a cell as described above, and a feeble electric current passed through it. Crystals of lead, in beautiful fern-like forms, rapidly deposit upon the negative electrode. On reversing the current the crystals will be seen to dissolve in the liquid, and while they are undergoing solution curious long filaments, resembling spiders' legs, make their appearance at the opposite electrode.

The experiment is best made in a small cell, using the microscope or the arrangement described on page 327. A current reverser should be placed in the circuit, so as to avoid disturbing the apparatus by changing the wires.

A solution of stannous chloride also deposits crystals of tin when submitted to treatment in the same way.

659. The deposition of certain metals in the crystalline form by displacement from solutions of their salts, by means of more electro-negative metals, may also be shown upon the screen.

The electrodes are replaced by a narrow strip of zinc about a millimetre wide, cut from a piece of thin sheet zinc. This is attached to a cork and introduced into a small cell and an image of it projected upon the screen. A solution of lead acetate, or stannous chloride, or silver nitrate, is then poured into the cell, and almost immediately crystals of the particular metal will begin to make their appearance upon the little strip of zinc, growing in the course of a few minutes to a considerable size.

660. Divided cells. In many electrolytic experiments it is necessary to divide the cell by means of a porous partition or diaphragm. This partition is most conveniently made of blotting-paper, by stitching several thicknesses of paper together. A supply of them may be made in the following way. Eight sheets of blotting-paper are first pinned together at the corners with paper fasteners, and a number of lines are ruled on the uppermost sheet. The distance between the lines should be just a little less than the width of the cell for which the parti-

Fig. 245.

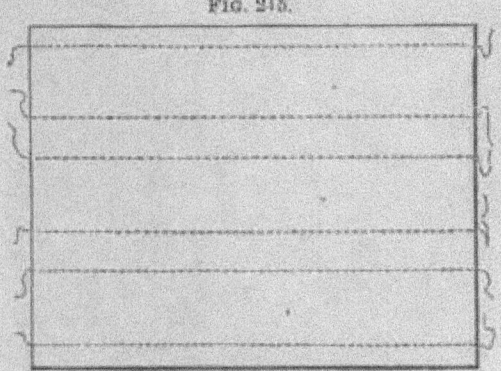

tion is to be used, and a space of about 6 millimetres should be allowed between each pair of lines. Then with a sewing-machine a row of stitches is run down each line, when the paper will present the appearance shown in fig. 215. With a pair of scissors the paper is then cut into strips, the cutting being made down the middle of the narrow spaces between each pair of lines.

A piece is cut from one of these strips, equal in length to the depth of the cell, and the edges opened out down the middle. It is then gently pushed down between the two walls of the cell, making a perfect fitting partition (A, fig. 216). A little margin for adjusting such a partition to a cell slightly narrower,

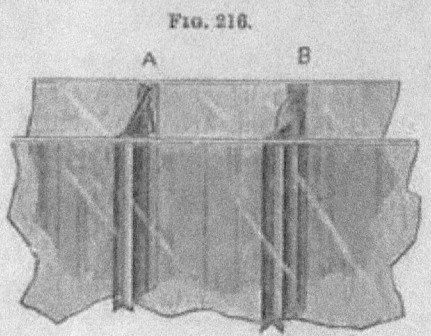

FIG. 216.

can be obtained by opening the paper between the rows of stitches, by gently pushing a lead pencil or glass rod down the middle. The partition so opened is shown at B, fig. 216.

661. In a divided cell as above described, the electrolysis of saline solutions can be shown, in which one or both of the final products of the decomposition can be rendered manifest by certain colour reactions, as for example :—

(a) **Cupric chloride, coloured with aniline blue.** When this solution is in a divided cell, the chlorine liberated at the anode will bleach the colouring matter in that side of the cell, leaving the pale green colour of the cupric chloride.

(b) **Potassium iodide, containing a little starch.** A small quantity of clear starch emulsion (see page 111) is added to a solution of potassium iodide, and the liquid electrolysed in a divided cell. The presence of the iodine will be instantly made evident by the liquid in the portion of the cell which contains the anode becoming blue.

(c) **Sodium chloride, coloured with reddened litmus.** In this case the liquid will be bleached in one division of the cell, and turned blue in the other.

(d) **Sodium sulphate, coloured with infusion of red cabbage.** One or two leaves of red cabbage are cut into shreds, and boiling water poured upon them in a beaker. The infusion so obtained has a violet colour. A saturated neutral solution of sodium sulphate is coloured with this solution, and the mixture placed in a divided cell. When electrolysis takes place the liquid in one portion of the cell becomes a bright red, while the alkali in the other division turns the liquid green.

662. To show that the products of electrolytic decomposition are only liberated in contact with the electrodes, and not in the intervening liquid, a cell divided into three compartments by means of two paper diaphragms can be used. Into a cell so divided sodium sulphate solution coloured with infusion of red cabbage is introduced, and the two electrodes are placed in the two extreme compartments. When the current is passed through the cell, the liquid in the central division remains unchanged, while that in the others is turned red in one case and green in the other.

663. By a modification of Experiment 661 (c) the principle of Castner's new electrolytic process for manufacturing caustic soda may be illustrated. For this purpose a cylindrical stoppered funnel is fitted with a cork, through which pass a small rod of gas carbon and an exit tube (fig. 217). Into the lower end of the funnel a platinum wire is fused. A quantity of mercury is placed in the apparatus, and upon this is poured a saturated solution of common salt. The carbon rod is made the anode, and the chlorine which is there disengaged may be collected. The sodium which is liberated at the mercury cathode is dissolved in that metal, forming sodium amalgam. During the electrolysis the apparatus must be gently shaken, in order to prevent the sodium amalgam from accumulating upon the sur-

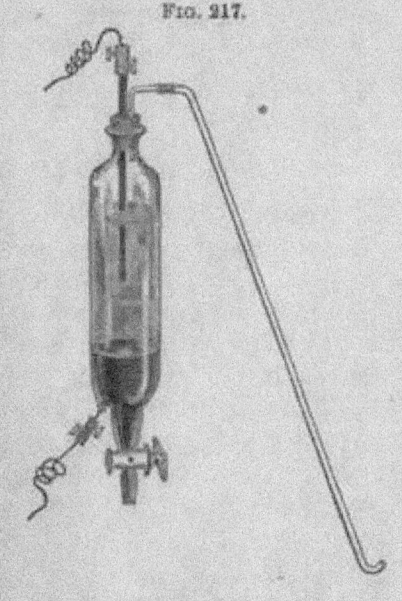

FIG. 217.

face of the metal and acting upon the water with the liberation of hydrogen. If this is done, practically no gas will be seen to rise from the surface of the mercury. After a few minutes—say when a moderate-sized cylinder of chlorine has been collected —the current may be stopped, and the mercury at once run out into a clean stoppered bottle. A little distilled water is then poured upon it and the bottle shaken for a moment, after which the water will be found to be strongly alkaline to test-paper.

664. As an illustration of the reduction of a metal from a fused electrolyte, the preparation of lithium from lithium chloride is a convenient example. The salt is carefully dried, and fused in a small porcelain crucible by means of a Bunsen flame, and a small rod of gas carbon is made the anode. A piece of stout iron wire, flattened out at one end, is used as the cathode. When the latter is introduced into the liquid, metallic lithium quickly collects on the iron. In a minute or two the wire is withdrawn and dipped into petroleum to cool it. The lithium may then be cut off the flattened wire with a knife.

LANTERN ILLUSTRATIONS

Not only are there a great many experiments which may with advantage be shown to an audience by means of the lamp, but there are others which can only be rendered visible by pro-jection upon a screen. The necessary apparatus for such lantern experiments as are described in the foregoing chapters, and the best methods for its disposition, may be described under the following heads—viz. the light, the lantern, and the lenses.

1. The source of light.—Practically this will be either the limelight or the electric light, although in the absence of these, and in a small room, for some experiments, a good duplex or triplex oil lamp can be made to do duty fairly well.

(a) The limelight can be produced by causing the flame of either burning hydrogen and oxygen, coal-gas and oxygen, or ether and oxygen to play upon a small cylinder of lime, and thereby raise it to incandescence; these different flames being called respectively the oxy-hydrogen, oxy-coal gas, and the oxy-ether flames.

The form of jet employed for the oxy-hydrogen flame is one in which the two gases mix in a small chamber, packed with wire gauze, and the mixture is inflamed at a platinum nozzle. Fig. 218 shows such a jet. The two gases are passed in at O and H; they mix in the chamber C, and pass from thence to the nozzle.

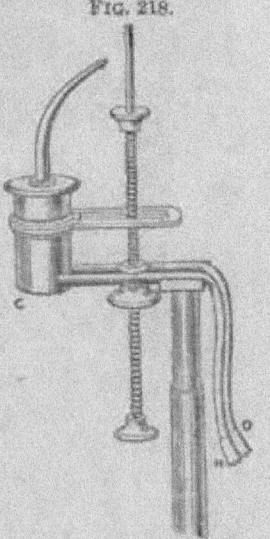

FIG. 218.

The oxy-hydrogen flame being very fine, the spot of light produced when it impinges upon the lime has the advantage of being smaller than that obtained by either the oxy-coal gas or oxy-ether flames; from the nature of this flame, however, it very rapidly bores little holes into the lime. This 'pitting' of the lime is attended with considerable risk to the nearest lens, especially when a short focus condenser is being used, for if the position of the lime be not frequently altered by a slight rotation of the cylinder, the flame will rebound from its surface and strike the lens. It is always advisable to guard the lens by interposing between it and the flame either a piece of sheet glass or thin mica.

(The clear sheets of mica used for covering photographs are well suited for this purpose.)

There are two kinds of limes usually supplied for the lime-light, known as 'hard' and 'soft;' for the oxy-hydrogen flame the hard limes are to be preferred.

The gas bags of former days have now been almost entirely superseded by iron or steel bottles containing the separate gases under pressure. In using these bottles, it is a great advantage to employ regulators with them. These are firmly screwed to the bottle, and reduce the pressure of the issuing gas to a few centimetres of mercury. Fig. 219 shows such a regulator. The

Fig. 219.

valve of the bottle may be fully opened, and the supply of gas controlled by the ordinary cocks upon the oxy-hydrogen jet, or by pinch-cocks upon the caoutchouc tube conveying the gas.

As these regulators are not intended to stand continuously the high pressure to which they are exposed, without a slight leakage, the valve of the bottle should be closed when the gas is not being used.

When starting the light, the hydrogen should be first turned on, and the jet inflamed. This should be allowed to play upon the lime for some little time, in order to warm it, before turning on the oxygen; as by so doing the lime is not so liable to be broken by the great heat. The hydrogen flame is then adjusted to a suitable size, and the oxygen very slowly turned on. The air in the tube, which has first to pass out, does not raise the temperature of the flame, and it is a common mistake to conclude that, as the light does not at once appear, therefore the stream of gas is not sufficiently rapid, and to open the oxygen cock more fully; the almost inevitable result of this is, that as soon as the oxygen reaches the flame there is a loud hissing and spluttering, and the flame is blown out. If the tube conveying the oxygen from the bottle to the jet be long, it is well to sweep the air out of it before attaching it to the jet by momentarily allowing oxygen to escape.

The oxy-coal gas flame is for many reasons more convenient than the oxy-hydrogen flame. The form of jet used when coal-gas is employed instead of hydrogen is known as the 'blow-through jet.' In this form of burner the gases are not mixed until they arrive at the jet, which consists of two concentric tubes, as in the ordinary Herapath blow-pipe. The coal-gas supplies the outer tube, the oxygen being fed into the flame by the inner tube. This flame has no tendency to pit the lime, and therefore does not endanger the lenses. With this arrangement 'soft' limes may be used, and, as in the case of hydrogen, it is well to warm the lime by allowing the coal-gas to burn against it for some time before passing the oxygen.

The oxy-ether flame is a very convenient source of light when neither hydrogen nor coal-gas are available, and gives an excellent light. The jet employed is similar to that for the oxy-hydrogen flame, but with a larger nozzle; each supply-pipe also should be furnished with a stop-cock. The special apparatus required is the ether tank, which consists of a copper box capable of holding about 800 cubic centimetres of ether.

At one end of the box is a 3-way tube, and at the opposite end a single exit tube, each provided with a stop-cock. The box is of special construction inside to insure the complete saturation with ether of the gas which passes through.

To use the apparatus the tank is first filled with ether, and the tube from the oxygen bottle attached to C, fig. 220. The oxygen bottle must be furnished with a regulator for this experiment. A tube from D is connected to the oxygen pipe of the jet, and a second from E to the hydrogen. The cocks A and B, as well as those upon the jet, are all closed, and the valve of the oxygen bottle opened. A and B are then opened, and the regula-

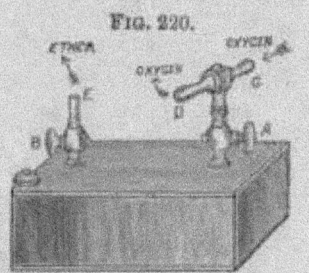

FIG. 220.

tion of the gases made entirely by means of the cocks upon the jet. The hydrogen cock is first opened and the gas ignited, and the size of flame regulated. (This flame is ether vapour with a little oxygen.) The oxygen tap is then gently turned until the required light is obtained. The regulation of the light must be done entirely by means of the taps upon the jet, and not those upon the tank, the real object of those being only to close the tank when not in use. At first sight it might appear a little dangerous to produce a mixture of ether vapour and oxygen in a closed vessel and convey it through a pipe to a flame, and there is no doubt that in the hands of a careless manipulator this apparatus might become dangerous; so long, however, as there is a good supply of ether in the tank (and the quantity it will hold will last many hours) there is no fear of the mixture exploding, and even should the tank run dry, the diminishing light gives ample warning of the fact long before the explosive mixture stage is reached, and then the light should be at once extinguished by closing both the cocks upon the jet. The light should never be extinguished by closing the valve of the oxygen bottle.

It is a good plan to place the oxygen bottle and the ether tank at a little distance from the lamp, to avoid all risk of falling into the mistake of touching any of the taps upon them, in the hurry of experimenting, and in the darkness.

For experiments extending over only a short time, when the light is not required for more than about fifteen minutes consecutively, an ether tank may be extemporised out of an ordinary two-necked Woulf's bottle, filled with tow, upon which sufficient ether is poured to thoroughly soak it. The bottle is fitted with two corks, one carrying a T tube, the long limb of which passes nearly to the bottom of the bottle, the other cork carrying a short exit tube. For complete safety the bottle may be either wrapped in a cloth, or put in a wooden box.

(b) The electric light. Although incandescent lamps have been designed for use in a lantern, the arc lamp is practically the only form of electric light that is available. When the current employed is obtained from a galvanic battery, the best form of lamp is the 'Foucault' lamp, in which the carbons are 'fed' by clockwork, and the arc maintained central by a double train of wheels (fig. 221).

When a dynamo current is to be used, the most convenient lamp is the newest form of the 'Brockie-Pell,' which has been designed especially for lantern work, and has been made small and compact in order to avoid the use of the enormous lanterns usually necessary to contain electric lamps. The carbons in this lamp are 'raked' in order to throw the light out horizontally, and the upper carbon also admits of adjustment so that

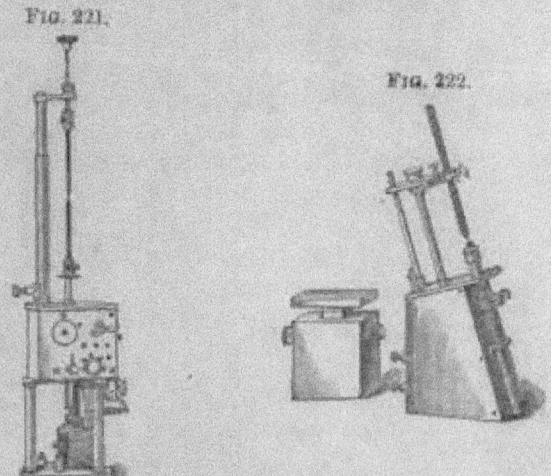

Fig. 221.

Fig. 222.

it can be fixed more or less behind the lower one (fig. 222). If run off accumulators this lamp burns absolutely noiselessly, but with a current direct from a dynamo it is liable to a slight hissing sound. This may be obviated by arranging a 'shunt' for the current, consisting of a number of cells made of strips of lead immersed in dilute sulphuric acid. The cells are preferably of glass, and about a pint capacity; the sheet lead should be bent as indicated in the figure. The number of cells used should be such that when the lamp is not burning there will only be a slight decomposition visible in the cells. A switch may be arranged to cut the current off the cells when not required. These cells when once set up require no further attention.

Fig. 223.

As most experiments only occupy a few minutes, a simple hand-lamp, in which the upper carbon is movable by a rack and pinion, may be used in the absence of a more elaborate instrument; in any case it is well to stand the lamp upon a small adjustable block, so that it may be raised or lowered at will (fig. 222). A range of about three centimetres is all the movement that is required.

II. The lantern.—As a rule, lanterns, especially for electric lamps, are made unnecessarily large. Where a portable lantern, chiefly for projecting slides, is required, none are better than those made on the model of the 'scyopticon.' These made-up instruments, however, are not well adapted for general lecture purposes where various experiments are to be shown on the screen; for this end it is better that the lantern should be quite a separate and isolated piece of apparatus. It should be as small

FIG. 224.

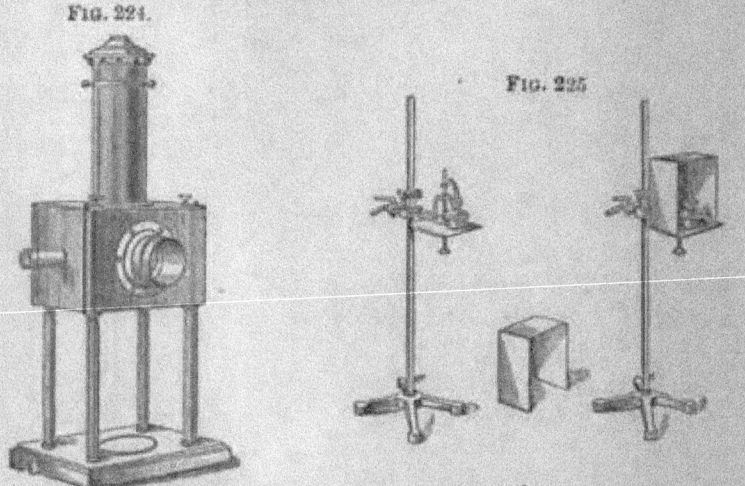

FIG. 225

as will conveniently surround the light; of such a height as to allow of plenty of room for fairly tall apparatus to be placed in front of it, and so made that stands and other objects can be brought close up to it. One of the most convenient models is the familiar 'Duboscq' lantern (fig. 224). It is not only suitable for the clockwork lamp for which it was designed, but can be used for the Brockie-Pell lamp above mentioned, and also for the limelight.

There are very few occasions, however, when an elaborate lantern is really indispensable, and for most purposes a sufficient screen for the lamp may easily be constructed of thin sheet iron. For the limelight this need only be a small hood attached to the burner, and projecting seven to ten centimetres in front of the lime. Fig. 225 shows such an arrangement; it is only necessary with such a lantern that the first lens should be provided with a screen, or flange, made of blackened cardboard, and of such a diameter that it intercepts all the light which escapes from the open front of the little hood.

III. The lenses.—For ordinary purposes of projection, when images of glass cells, tubes, and similar objects, as well as photographs and slides, have to be projected upon the screen, a converging beam should be used. For this purpose the following combination of lenses will be found to give good results:—

A. A double convex lens. 10 centimetres (four inches) diameter, and having a focal length of about 34 centimetres. (Such a lens as an ordinary reading glass.)

B. A plano-convex lens. 10 centimetres diameter and about 18 centimetres focal length.

C. A compound objective. 5 centimetres diameter and about 18 centimetres focal length. (An opera-glass objective makes an excellent lens for this purpose.)

Fig. 226 shows the arrangement of these lenses. The distance between the point of light L and the lens A with such a com-

bination will be about 12 centimetres, between A and B about 7 centimetres, and between B and C about 23 centimetres, thus allowing ample room for moderately bulky pieces of apparatus. Each of these lenses should be supported on its own independent stand, and the stands of A and B should be so made as to allow of the lenses being brought near to one another; each lens also should be surrounded with a flange or screen of black cardboard.

Some objects, of which it is desired to project an image, such as an ordinary lantern-slide, can be mounted in an inverted position, in order that the image may appear normal on the screen; in many cases, however, this is impossible, so that if the image is to appear the right way up it is necessary to employ a reversing prism. This consists of an equilateral right-angled glass prism (which need not measure more than 7 centimetres along the hypotenuse), supported immediately beyond the focussing lens c, fig. 226. This prism should be stood upon a small levelling block, and placed so that it receives the entire ray of light. Besides its use in re-inverting the image, it will be found of the greatest advantage for other reasons; it enables the experimenter to raise or lower the illuminated disc by means of the levelling block, without tilting or in any way disarranging

FIG. 227.

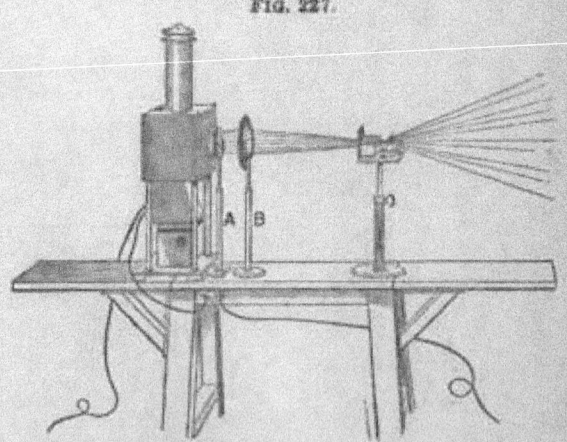

the lantern or any of the apparatus. All the apparatus can be placed at a level which is found most convenient for manipulating, and the image thrown up to any desired height by adjusting the angle of the prism. In the same way the image of any object, which from its nature it is difficult or impossible to support quite vertically, may be made to appear vertical upon the screen by a slight adjustment of the prism. Fig. 227 shows the general arrangement of the apparatus, the flange having been removed from lens A.

In the absence of a prism an image may be re-inverted upon the screen by means of two mirrors. The first is placed at an angle of 40° to the beam, so as to reflect it vertically upwards. The beam is then received upon the second mirror and thrown right back over the lantern, which in this arrangement must be placed with its back towards the screen. If the lantern is only turned half-way round, so that the second mirror diverts the beam in a direction at right angles to that which it has when reaching the first mirror, then the image will only be half inverted—that is to say, the image of a vertical object will appear horizontal, and *vice versa*. If common looking-glass is used in this arrangement the image loses somewhat in sharpness by reflection from both surfaces; this is not the case, however, if the mirrors are made by depositing silver upon glass,[1] and using the exposed and polished metal surface as the reflector.

[1] The best method for silvering glass is the following :—

Solution A.—Dissolve 90 grams sugar-candy in water; add

4 c.c. nitric acid (sp. gr. 1·22),

175 c.c. alcohol.

Make up to 1 litre with water. *This solution will keep.*

Solution B.—Dissolve 1·8 grams silver nitrate in 180 c.c. water; add ammonia drop by drop until the precipitate *nearly* redissolves; add 0·9 grams caustic potash dissolved in a few c.c. of water, and again nearly redissolve the precipitate by a few drops of ammonia. *This solution must be made up as required.*

For use, 10 c.c. Solution A. are mixed with 180 c.c. Solution B.

The glass to be silvered, after being thoroughly cleaned, is immersed for a moment or two in a solution of stannous chloride and then rinsed with water. It is then placed in the bath or dish containing the silvering mixture, and in about ten minutes the silvering is complete. The mirror is then rinsed with water and lightly rubbed over with a soft rag. By this process the silver is deposited very hard and bright.

Convenient dishes or trays in which to conduct the silvering operation are

Objects which are too bulky, or which for any other reason cannot conveniently be shown by means of this combination, may be placed in a wide parallel beam obtained by drawing lens B nearer to the light, and focussing the image upon the screen by means of lens A. The arrangement is seen in fig. 228.

Fig. 228.

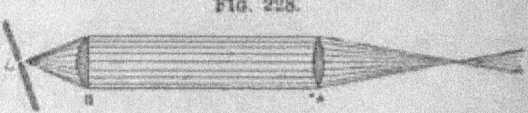

When it is required to project an image of objects in horizontal position, such as flat cells containing liquids, which

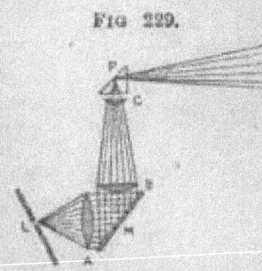

Fig 229.

must be retained in this position, the lenses A, B, C and the prism may be arranged in the position shown in fig. 229. The light, after passing through lens A, is received upon a piece of ordinary mirror supported at an angle of 45°, and so reflected upwards through lens B. After passing lens C the beam is again reflected at right angles by the prism.

Fig. 230 represents a convenient form in which the horizontal projector is made. The mirror is contained within the metal cylinder æ. It is furnished with lenses corresponding to B and C in the foregoing arrangements, and with a prism F. This latter is capable of three movements. It can be raised and lowered for focussing purposes by the rack and

made in the following way. A quantity of white paraffin wax is melted in any convenient vessel, and is brushed over a sheet of cartridge paper with a broad flat brush. When cold, a piece of this waxed paper is folded up into a shallow tray just a little larger than the glass to be silvered. The corners are folded over *without being cut*, and stuck down by gently warming the wax with a hot knife, or by the momentary application of a lighted taper. Scarcely any deposition of silver takes place upon such a tray, the whole of it going to the glass to be silvered, whereas in a glass or porcelain dish the silver deposits itself upon the vessel as well as the object.

pinion R; it can be tilted up by means of the screw s, in order
to place the luminous disc at any convenient height upon the
screen, and it can be turned upon its axis so that
the image may be thrown upon a screen placed
at any angle to it. The apparatus is used in
conjunction with lens A, as shown in the diagram
(fig. 229).

A short-focus condenser (about 6 centimetres
focal length) will be found a useful lens for giving
a narrow powerful beam of nearly parallel rays.

If a higher degree of magnification be required than is given by lens C in the above
arrangements, it may be obtained by the use of
an objective of shorter focus, such as an opera-
glass objective of about 10 centimetres focal
length.

The lantern microscope is an independent piece
of apparatus, which must be either adapted to fit the lantern
or firmly mounted upon its own stand, for as the whole light
has to converge upon such a small surface as the face of the
objective, it is absolutely necessary that there should be no
possibility for any shifting of the parts. The microscope is used
with a converging beam from lenses A and B, or from a short
focus condenser.

For most experiments, however, where a lantern microscope
is required (such as Nos. 77 and 112) the following simple
arrangement may be substituted. By means of a short-focus
condenser the beam is made to converge to a small spot of light
upon the object, and an image projected by means of an eye-
piece taken from an ordinary microscope. The eye-piece is
supported upon a small wooden V-block, the beam being sent
through it in the same direction as one looks through when
using it on the microscope. A flange of blackened cardboard
is placed on the end of the eye-piece in order to intercept any
scattered light. This arrangement is very suitable for showing
the process of crystallisation upon the screen (see Experiment

FIG. 230.

112) or for microscope slides not requiring very high magnification, such as crystals of various salts. For example, the different crystalline character of similarly constituted salts, as potassium and sodium nitrates, may readily be shown by allowing a drop of solutions of these compounds to slowly evaporate upon a microscope slide. The crystals so obtained may be covered, in order to preserve the slide, by surrounding them with a small flat india-rubber ring, which is made to adhere to the glass by means of a touch of gum or balsam, and then placing upon this a microscope cover-glass.

APPENDIX.

TABLE I.

Table showing the concentration of oxygen by repeated solution of air in water.

Composition of air—

Oxygen	20·96
Nitrogen	79·04
	100·00

Composition of air dissolved by water—

Oxygen	34·91
Nitrogen	65·09
	100·00

Composition of air after second solution in water—

Oxygen	47·5
Nitrogen	52·5
	100·0

Composition of air after third solution—

Oxygen	75
Nitrogen	25
	100

TABLE II.

Table showing the difference between the solubility in water of pure oxygen, and of oxygen mixed with nitrogen, as in the atmosphere.

1 litre of water dissolves :—

At					From pure oxygen	From air
0°C.	.	.	.	.	48·9 c.c.	10·2 c.c.
15°C.	.	.	.	.	34·1 c.c.	7·1 c.c.
25°C.	.	.	.	.	28·4 c.c.	5·8 c.c.

TABLE III.

Table showing the change of volume of water on being heated from 0° C. to 8° C.

1·000000 volume of water at	0°C. becomes
0·999915	"	"	+ 2°C. "
0·999870	"	"	4°C. "
0·999900	"	"	6°C. "
1·000000	"	"	8°C. "

TABLE IV.

Solubility of various salts in water.

Salt	Weight in grams dissolved by 100 grams of water at—		
	0°C.	20°C.	100°C.
Sodium chloride	35·7	36·0	39·7
Potassium chloride	29·2	35·0	57·0
Ammonium chloride	28·4	37·3	73·0
Barium chloride	—	42·2	72·0
Potassium bromide	53·4	64·5	102·0
Potassium iodide	126·0	143·0	200·0
Potassium chlorate	3·33	7·2	59·8
Potassium nitrate	13·3	31·7	246·0
Sodium nitrate	66·7	84·9	168·0
Calcium sulphate, 2 aq.	0·2	0·25	0·5
Calcium sulphate (Anhyd.)	0·205	0·241	0·217
Magnesium sulphate (Cryst.)	26·23	35·32	73·27
Copper sulphate (Cryst.)	31·61	42·31	203·32
Zinc sulphate (Cryst.)	115·22	161·50	653·6
Potassium sulphate	8·36	10·6	26·0
Potash alum (Cryst.)	3·3	15·4	357·5
Borax (Cryst.)	2·83	7·88	about 200·0
Potassium bichromate	5·0	13·0	102·0
Sodium sulphate (Anhyd.)	4·5	20·0	43·0
Sodium sulphate (Cryst.)	12·16	58·35	212·47

	0°C.	20°C.	34°C.	37·6°C.	40°C.	100°C.
Sodium sulphate (Cryst.)	12·16	58·35	112·22	355·0	324·9	212·47

TABLE V.

Boiling-points of saturated solutions of various salts in water.

	°C.	Grams of salt to 100 grams of water
Sodium carbonate	101·6	48·5
Sodium chloride	108·4	41·2
Potassium nitrate	115·9	335·1
Sodium nitrate	121·0	224·8
Potassium carbonate	133·0	205·0
Calcium chloride	179·5	325·0

TABLE VI.

Tension of aqueous vapour.

Temperature, degrees Centigrade	Tension in millimetres of mercury	Temperature, degrees Centigrade	Tension in atmospheres
°		°	At.
−20	0·927	100	1
−10	2·093	111·7	1·5
0	4·600	120·6	2
+5	6·534	127·8	2·5
10	9·165	133·9	3
15	12·699	144·0	4
20	17·391	159·2	6
30	31·548	170·8	8
40	54·906	180·3	10
50	91·982	188·4	12
60	148·791	195·5	14
70	233·093	201·9	16
80	354·280	207·7	18
90	525·450	213·0	20
100	760·000	224·7	25

TABLE VII.

Solubility of gases in water.

1 vol. of water at 0°C. and under a pressure of 760 mm. will dissolve of—

	Vols. measured at 0°C. and 760 mm.
Ammonia	1148·00
Hydrochloric acid	503·00
Sulphur dioxide	79·789
Sulphuretted hydrogen	4·73
Carbon dioxide	1·7967

	Vols. measured at 0°C. and 760 mm.
Nitrous oxide	1·3052
Ethylene	0·25629
Marsh gas	0·05449
Oxygen	0·04114
Carbon monoxide	0·03287
Air	0·02471
Nitrogen	0·020346
Hydrogen	0·0193

Chlorine,[1] coefficient of absorption at 10°C. = 3·0361.

TABLE VIII.

Solubility of hydrochloric acid in water at various temperatures

1 c.c. of water at 760 mm. dissolves

At	Grams HCl	c.c. at 0°C. and 760 mm
0°C. . . .	0·825 . . .	503
4° . . .	0·804 . . .	491
8 . . .	0·783 . . .	479
12 . . .	0·762 . . .	466
16 . . .	0·742 . . .	453
20 . . .	0·721 . . .	440
30 . . .	0·673 . . .	411
40 . . .	0·633 . . .	387
50 . . .	0·596 . . .	364

TABLE IX.

Solubility of ammonia in water at various temperatures.

1 c.c. of water at 760 mm. dissolves

At	Grams NH₄	c.c. at 0°C. and 760 mm
0°C. . .	0·875 . . .	1148
4 . . .	0·792 . . .	1040
8 . . .	0·713 . . .	923
12 . . .	0·645 . . .	845
16 . . .	0·582 . . .	764
20 . . .	0·526 . . .	690
30 . . .	0·403 . . .	529
40 . . .	0·307 . . .	403
50 . . .	0·229 . . .	308

[1] Below 10°C. chlorine forms a solid hydrate with water.

TABLE X.

Solubility of sulphur dioxide in water at various temperatures.

1 c.c. water at 760 mm. dissolves

At 0°C.	c.c. SO₂ at 0°C. and 760 mm.
0	79·789
4	69·820
8	60·805
12	52·723
16	45·578
20	39·374
30	27·161
40	18·766

TABLE XI.

Solubility of carbon dioxide and nitrous oxide in water at various temperatures.

1 c.c. water at 760 mm. dissolves

At 0°C.	c.c. CO₂ at 0°C. and 760 mm.	c.c. N₂O at 0°C. and 760 mm.
0	1·7967 . .	1·3052
5	1·4497 . .	1·0954
10 . . .	1·1847 . .	0·9196
15 . . .	1·0020 . .	0·7778
20 . . .	0·9014 . .	0·6700
25 . . .	— . .	0·5962

TABLE XII.

Classification of waters in order of their softness.

1. Rain water.
2. Upland surface water.
3. Surface water from cultivated land.
4. Polluted river water.
5. Deep well water.
6. Shallow well water.

TABLE XIII.

Soap destroyed by 100,000 lbs. of various waters.

	lbs.
Thames water	212
River Lea	204
Kent Co.'s water	256

		lbs.
South Essex Co.'s water	.	253
Water supply of Leicester	.	161
" " " Preston	.	80
" " " Manchester	.	32
" " " Glasgow (L. Katrine)	.	4
Bala Lake	.	5
Thirlmere	.	8

TABLE XIV.

Weight and cost of different materials required to soften the same quantity of water.

	£	s.	d.
1 cwt. quicklime	0	0	8
4¾ cwts. sodium carbonate	2	17	9
20½ cwts. soap	47	1	8

TABLE XV.

Composition of sea water (Irish Sea).

Water	966·14054
Saline matter consisting of:—	
Sodium chloride	26·43918
Potassium chloride	0·74619
Magnesium bromide	0·07052
Magnesium chloride	3·15083
Magnesium sulphate	2·06608
Magnesium carbonate	traces
Calcium sulphate	1·33158
Calcium carbonate	0·04754
Lithium chloride	traces
Ammonium chloride	0·00044
Magnesium nitrate	0·00207
Silicon dioxide	traces
Ferrous carbonate	0·00503
	33·85946
	1000·00000

TABLE XVI.

Analysis of typical good and bad natural waters.

	No. 1. Good water		No. 2. Bad water	
—	Parts per million	Grains per gallon	Parts per million	Grains per gallon
Total solids	63·0	4·4	530·0	37·1
Nitrogen as nitrates and nitrites .	0·25	0·017	7·8	0·546
Free ammonia	0·03	0·002	4·32	0·303
Albuminoid ammonia . . .	0·07	0·005	0·9	0·063
Chlorine	11·4	0·8	69·0	4·8
Temporary hardness . . .	—	0·1	—	7·2
Permanent hardness . . .	—	2·4	—	14·4
Total hardness	—	2·5	—	21·6

TABLE XVII.

Composition of boiler incrustations.—(Macadam).

I. CARBONATE CLASS.

District from which the samples were obtained	Dunbar	Selkirk	Slough	Edinburgh	Carlisle
Ferric oxide (Fe₂O₃) } Aluminic oxide (Al₂O₃) }	7·46	2·96	2·36	2·48	2·96
Calcium carbonate (CaCO₃) .	32·16	74·25	50·04	69·95	75·92
Calcium sulphate (CaSO₄) .	5·64	3·08	29·76	20·50	3·16
Magnesium carbonate (MgCO₃)	20·04	3·76	10·84	7·24	10·16
Sodium salts	3·31	1·15	0·86	0·85	0·84
Silica (SiO₂)	16·94	8·56	4·28	3·76	4·94
Organic matter	7·70	3·02	0·48	0·12	0·22
Moisture	6·78	3·10	1·22	1·22	1·53
Total .	110·03	99·88	99·84	99·43	99·73

TABLE XVIII.

II. SULPHATE CLASS.

District	Preston Pans	Granton	Slough	Smeaton	Carlisle
Ferric oxide . . } Aluminic oxide . }	4·64	3·56	5·04	1·68	4·06
Calcium carbonate . .	1·22	26·56	25·62	9·72	1·21
Calcium sulphate . .	78·32	38·16	55·92	55·28	50·36
Magnesium carbonate .	10·36	23·16	5·56	18·40	5·60
Sodium salts . . .	0·64	1·75	0·22	0·56	0·21
Silica	3·22	5·42	5·26	9·46	36·22
Organic matter . .	0·56	0·12	0·36	2·64	0·68
Moisture . . .	0·72	1·04	1·34	1·67	1·88
Total . .	99·68	99·77	99·32	100·41	100·22

TABLE XIX.

Table of the specific heat of the elements in the solid state.

Element	Atomic weight	Specific heat	Atomic heat	Weights containing equal quantities of heat
Lithium	7	0·94	6·6	7
Boron	11	0·5	5·5	13·2
Carbon	12	0·46	5·5	14·3
Sodium	23	0·29	6·7	22·7
Magnesium	24·4	0·25	6·1	26·3
Aluminium	27	0·21	5·7	31·8
Silicon	28·2	0·20	5·6	32·9
Phosphorus	31	0·17	5·3	38·7
Sulphur	32	0·16	5·1	41·1
Potassium	39	0·17	6·6	38·7
Calcium	40	0·17	6·8	38·7
Manganese	55	0·12	6·6	54·8
Iron	56	0·11	6·2	59·7
Nickel	58·6	0·11	6·4	59·7
Copper	63·2	0·094	5·9	70·0
Zinc	65·3	0·094	6·1	70·0
Arsenic	75	0·081	6·1	81·2
Silver	107·7	0·056	6·0	118
Tin	118	0·056	6·6	118
Antimony	120	0·051	6·1	129
Iodine	127	0·054	6·9	122
Platinum	194·4	0·033	6·4	199
Gold	196	0·032	6·3	205
Mercury	200	0·032	6·4	206
Lead	206·5	0·031	6·4	212
Bismuth	208·2	0·031	6·4	212

TABLE XX.

Diffusion of gases. (Graham, 1834.)

Gas	Density (air=1)	Square root of density	$\frac{1}{\sqrt{\text{Density}}}$	Actual velocity of diffusion by experiment
Hydrogen	0·06926	0·2632	3·7794	3·83
Marsh gas	0·559	0·7476	1·3375	1·344
Steam	0·6235	0·7896	1·2664	—
Carbon monoxide	0·9678	0·9837	1·0165	1·1149
Nitrogen	0·9713	0·9856	1·0147	1·0143
Ethylene	0·978	0·9889	1·0112	1·0191
Nitric oxide	1·039	1·1096	0·9808	—
Oxygen	1·1056	1·0515	0·9510	0·9487
Sulphuretted hydrogen	1·1912	1·0914	0·9162	0·95
Nitrous oxide	1·527	1·2357	0·8092	0·82
Carbon dioxide	1·52901	1·2365	0·8087	0·812
Sulphur dioxide	2·247	1·4991	0·6671	0·68

TABLE XXI.

Composition of air.

	Volumes per 1,000
Nitrogen (containing about 1 per cent. of argon) .	779·0600
Oxygen	206·5940
Aqueous vapour	14·0000
Carbon dioxide	0·3360
Ammonia	0·0080
Ozone	0·0015
Nitric acid	0·0005
	1000·0000

TABLE XXII

Composition of air from various localities in 100 parts by volume.

	Oxygen. Parts by volume	Nitrogen. Parts by volume
St. Bartholomew's Hospital . .	20·885	79·115
	20·999	79·001
Paris	20·913	79·087
	20·999	79·001
Lyons	20·918	79·082
	20·966	79·034
Toulon	20·912	79·088
	20·982	79·018
Berlin	20·908	79·092
	20·998	79·002
Geneva	20·909	79·091
	20·993	79·007
Montanvert	20·963	79·037
Summit of Pichincha, 16,000 feet .	20·949	79·051
	20·988	79·012
North American prairie . .	20·910	79·090
South America . . .	20·960	79·040
Liverpool to Vera Cruz . .	20·918	79·082
	20·965	79·035
18,000 feet above London .	20·885	79·115
Manchester . . .	20·876	79·124
	20·888	79·112

Z

TABLE XXIII.

Iodine in seaweed.

Dry weeds		Per cent.	lbs. Per ton
Drift kelp:—			
Laminaria digitata. Tangle. Stem	.	0·4535	10·158
„ „ „ Frond	.	0·2946	6·593
„ „ stenophylla	.	0·4777	10·702
„ „ saccharina. Sugar wrack	.	0·2794	6·258
Cut kelp:—			
Fucus serratus. Black wrack	.	0·0856	1·807
„ vesiculosus. Bladder wrack	.	0·0297	0·665
Ascophyllum nodosum. Knobbed wrack	.	0·0572	1·281
Various :—			
Halidrys siliquosa. Sea oak	.	0·2131	4·773
Japanese seaweed (edible)	.	0·3171	7·102
Himanthalia lorea. Sea laces	.	0·0892	1·998
Rhodymenia palmata. Dulse (edible)	.	0·0712	1·594
Chorda filum. Sea twine	.	0·1200	2·688
Zostera marina. Grass wrack	.	0·0457	1·023
D'Urvillea utilis (Falkland Islands)	.	0·0075	0·179
Macrocystis pyrifera	.	0·0308	0·690

TABLE XXIV.

Analyses of kelp.

—	Kelp		Ash
	Irish	Scotch	Scotch
Potassium sulphate	11·14	13·95	12·71
Potassium chloride	27·17	17·79	18·09
Sodium chloride	9·00	14·00	6·80
„ carbonate	5·82	3·92	3·43
„ sulphide	Heavy traces	Heavy traces	Slight traces
„ thiosulphate	1·22	0·75	0·17
„ iodide	0·82	0·76	1·48
„ sulphocyanide	Heavy traces	Heavy traces	Slight traces
Soluble organic matter	0·00	0·00	0·42
Insoluble organic matter	41·41	44·80	49·75
Water	3·10	4·05	7·00
	99·68	100·02	99·85
Total potash	23·17	18·77	18·32
Iodine, lbs. per ton	15·5	14·5	28·0
Carbon in insoluble part	Nil	Nil	2·0

TABLE XXV.

Dissociation of iodine molecules.

Temperature	Specific gravity	Percentage decomposition	Rise of temperature	Increase in percentage decomposition	Mean increase in decomposition for 100°
448°	8·74	—	—	—	—
680	8·23	—	—	—	—
760	8·28	—	—	—	—
855	8·07	8·6	—	—	—
940	7·60	14·5	85°	5·9	6·9
1043	7·01	25·0	103	10·5	10·2
1275	5·82	50·5	232	25·5	11·0
1390	5·27	66·2	115	15·7	13·7
1468	5·06	73·1	78	6·9	8·8

TABLE XXVI.

Dissociation of nitrogen peroxide.

Temperature	Density, hydrogen=1	Percentage decomposition
26·7°C.	38·3	20·00
60·2	30·1	52·04
100·1	24·3	89·23
135·0	23·1	98·69
140·0	22·96	100·00

TABLE XXVII.

Gases absorbed by charcoal (Hunter).

One volume of charcoal (from cocoa-nut shell) absorbs of—

	Vols. at 0°C. and 760 mm.
Ammonia	171·7
Cyanogen	107·5
Nitric oxide [1]	86·3
Ethylene	74·7
Nitrous oxide	70·5
Phosphoretted hydrogen	69·1

[1] As a broad rule the more easily condensible the gas, the more it is absorbed by charcoal. Nitric oxide appears to shew an exception, as it is a much more difficultly condensible gas than nitrous oxide. As these results of Hunter's were published as far back as 1865 (*Phil. Mag.*, 4, xxix. p. 116), it may be doubted whether the gas he experimented upon was pure nitric oxide, as at that date the true action of nitric acid upon copper was not known.

	Vols. at 0°C. and 760 mm.
Carbon dioxide	67·7
Carbon monoxide	21·2
Oxygen	17·9
Nitrogen	15·2
Hydrogen	4·4

TABLE XXVIII

One volume of beech-wood charcoal absorbs (Saussure) of—

	Vols. at 12°C. and 724 mm.
Ammonia	90
Hydrochloric acid	85
Sulphur dioxide	65
Sulphuretted hydrogen . . .	55
Nitrous oxide	40
Carbon dioxide	35
Ethylene	35
Carbon monoxide	9·42
Oxygen	9·25
Nitrogen	6·50
Hydrogen	1·20

TABLE XXIX.

Table of strength of aqueous sulphuric acid.

Specific gravity at 15°C.	Percentage of H_2SO_4.	Specific gravity at 15°C.	Percentage of H_2SO_4.
1·006	0·9	1·308	40·2
1·014	2·8	1·357	45·5
1·029	4·8	1·410	51·2
1·045	6·8	1·453	55·4
1·060	8·8	1·498	52·6
1·075	10·8	1·563	65·5
1·091	13·0	1·615	70·0
1·108	15·2	1·671	74·7
1·134	18·5	1·732	79·9
1·152	20·8	1·774	84·1
1·190	25·8	1·796	86·5
1·220	29·6	1·819	89·7
1·274	36·0	1·842	100·0

TABLE XXX.

Increase of light by heating air and coal gas before combustion.[1]

		Rate of consumption per hour	Light in sperm candles each burning 120 grains per hour
1. Argand without external chimney	{	3·3 cubic feet	13·0 candles
		3·7 „	15·5 „
		4·2 „	17·0 „
		2·2 „	13·0 „
2. Same argand with external chimney	{	2·6 „	15·5 „
		2·7 „	16·7 „
		3·0 „	19·7 „
		3·3 „	21·7 „

Gain with equal light . . . . = 49 per cent.
Gain with equal consumption . . . = 67 „ „

TABLE XXXI.

Critical temperature of gases.[2]

	°C.
Hydrogen	−238
Nitrogen	−149
Carbon monoxide	−136
Oxygen	−118·8
Nitric oxide	− 93·5
Marsh gas	− 82
Ethylene	+ 9
Carbon dioxide	31·35
Acetylene	35
Nitrous oxide	37
Hydrochloric acid	52·3
Ammonia	131
Chlorine	141
Sulphur dioxide	155·4
Water	370·0

TABLE XXXII.

Liquefaction of gases.

	C.			
Sulphur dioxide at	{ 0 and under a pressure of	1·53 atmospheres
	+ 15 „ „	3·0 „
	− 10 „ „	ordinary pressure
Hydriodic acid .	0 „ „	4·0 atmospheres

[1] These results were obtained by the use of an argand lamp provided with an outer shade or chimney, by means of which all the air used for the combustion is made to brush past the heated inner chimney. In this way the air becomes very strongly heated, and this hot air in playing over the portions of the burner raises the temperature of the metal, and consequently of the issuing coal gas.

[2] The critical temperature of a gas is that point above which if the gas be raised no pressure will be able to reduce it to the liquid condition.

	°C.				
Ammonia	$\{$	0	and under a pressure of	4·2	atmospheres
		15·5	,,	6·9	,,
Chlorine	$\{$	0	,,	6·0	,,
		+ 12	,,	8·5	,,
		− 34	,,	ordinary pressure	
Hydrochloric acid	$\{$	− 16	,,	20	atmospheres
		− 4	,,	25	,,
		+ 10	,,	40	,,
Nitrous oxide		0	,,	30	,,
Ethylene	$\{$	0	,,	41	,,
		− 75	,,	5	,,
		−103	,,	ordinary pressure	
Acetylene	$\{$	+ 1	,,	48	atmospheres
		+ 10	,,	65	,,
Carbon dioxide	$\{$	− 5	,,	30·8	,,
		+ 5	,,	40·4	,,
		+ 15	,,	52·1	,,
Nitric oxide	$\{$	− 93·5 (critical temp.)	,,	71·2	,,
		− 97·5	,,	57·8	,,
		−105·0	,,	41·0	,,
Marsh gas		− 82·0 (critical temp.)	,,	55·8	,,
Oxygen	$\{$	−118·8 (critical temp.)	,,	58·0	,,
		−129·6	,,	27·0	,,
Nitrogen	$\{$	−149·0 (critical temp.)	,,	27·5	,,
		−148·2	,,	31·0	,,

TABLE XXXIII.

THE PERIODIC CLASSIFICATION OF THE ELEMENTS.

	Group 0, the Inert Gases	Group I		Group II		Group III		Group IV		Group V		Group VI		Group VII		Group VIII
		A	B	A	B	A	B	A	B	A	B	A	B	A	B	
1st short, or typical, period	He	Li	(H)?	Be			B		C		N		O	(H)?	F	
2nd short period	Ne	Na		Mg			Al		Si		P		S		Cl	
1st long period — even series	A	K		Ca		Sc		Ti		V		Cr		Mn		Fe, Co, Ni
1st long period — odd			Cu		Zn		Ga		Ge		As		Se		Br	
2nd long period — even series	Kr	Rb		Sr		Y		Zr		Nb		Mo				Ru, Rh, Pd
2nd long period — odd			Ag		Cd		In		Sn		Sb		Te		I	
3rd long period — even series	X	Cs		Ba		La		Ce								
3rd long period — odd			Au		Hg		Tl		Pb		Bi					
4th long period — even series						Yb				Ta		W				Os, Ir, Pt
4th long period — odd																
5th long period — even series						Th						U				
5th long period — odd																
Oxygen compounds	—	R_2O		R_2O_2		R_2O_3		R_2O_4		R_2O_5		R_2O_6		R_2O_7		(RO_4)
Hydrogen compounds	—	—		—		—		RH_4		RH_3		RH_2		RH		—

INDEX

ABS

ABSORPTION of gases by charcoal, 182, 339
Absorption of gases by water, 71, 331
Acetylene, 212
Air (See Atmosphere), 154
— composition of, 159, 337
Ammonia, 128
— combustion in oxygen, 129
— — — chlorine, 130
— combustion of potassium in, 133
— liquefaction of, 134, 294
— solubility in water, 133, 332
— solution of potassium in liquid, 135
Ammonium amalgam, 137
— chloride, 136
— — dissociation of, 287
Antimoniuretted hydrogen, 285
Antimony, amorphous, 283
— combustion of, 112, 284
Argon, 126
Arsenic, 281
Arseniuretted hydrogen, 282
Atmolysis, 166
Atmosphere, 154
— carbon dioxide in, 186
— formation of fogs in, 156
— suspended matter in, 154
— volume composition of, 118, 159
— weight of, 158
Azoimide, 138

Balance, construction of, 8
Boiler, 3
— incrustation, composition of, 335
Boiling-points of saturated solutions of salts, 331
Boric acid, 241
Boron, 240
Bromine, 105
— action of, on metals, 105

CUP

Bromine vapour, absorption of, by charcoal, 182

Cailletet apparatus, 303
Calcium phosphide, action of water upon, 254
Carbon, 179
— dioxide, 186
— — combustion of metals in, 191
— — critical point of, 307
— — solidified, 194
— — solubility in water, *table*, 333
— disulphide, 276
— — combustion of potassium in, 277
— — decomposition of, by detonation, 279
— monoxide, 197
Charcoal, absorption of gases by, 182, 294
— — — — — table of, 339
— decolourising power of, 186
— gases in, 184
Chloric acid, 103
Chlorine, 79
— action on metals, 84
— behaviour of dry, 86
— combination with hydrogen, 89
— monoxide, 100
— oxides and acids of, 100
— peroxide, 100
— preparation of liquid, 83
Collodion balloons, 11
Combustion, 167
— increase of weight resulting from 171
Connectors, 3
Critical temperature of gases, 341
Cryophorus, 62
Cuprous acetylide, 217
— chloride, 213

DEA

Deacon's process, 82
Diamond, combustion of in oxygen, 179
Diffusion of gases, 161; 336
— through cracked vessel, 162
— — open tube, 162
— — porous material, 163
— — soap film, 165
Dissociation, 286

Electric lamps, 320
Electrolysis, experiments on, 311
Ethylene, 207
Ether tank, 319
Euchlorine, 101, 265

Fire syringe, 298
Flame, luminosity of, 217
— the Bunsen, 228
— vortices, 218
Flames, shadows cast by, 226
— solid matter in, 217
Fluorine, 114
Fulminate, mercuric, 280

Gaseous diffusion, 161
Gases, absorption by charcoal, 182, 294
— collection and keeping of, 81
— solubility in water, 71, 331
Graham's diffusiometer, 163
Graphite, combustion of, in oxygen, 180

Holmes signal, 254
Hydrazoic acid, 138
Hydriodic acid, 113
Hydrobromic acid, 106
— — synthetical formation, 108
Hydrochloric acid, 94
— — electrolysis of, 95
— — formation by synthesis, 95
— — solubility in water, 99, 332
Hydrofluoric acid, 115
Hydrogen, 1
Hydrogen peroxide, 74
Hydrogen persulphide, 267
Hydrogenium, 18
Hypochlorous acid, 102

NIT

Ice flowers, 64
Ignition point, 175
Increase of light by heating air and coal-gas, 341
Iodic acid, 113
Iodine, 110
— in seaweed, 338
— oxides and oxyacids of, 113
Iron pyrophorus, 176

Kelp, analyses of, 338

Lantern illustrations, 316
— microscope, 327
Lanterns, 322
Lenses, 323
Levelling table, 60
Limelight, 316
Liquefaction of gases, 293
— — — table of, 341
Lithium nitride, 126
Luminosity of flame, 217

Magnesium, action of, upon steam, 2
— amalgam, 4
— combustion of, in carbon dioxide, 192
— flash light, 27, 92
— nitride, 124
Marsh gas (Methane), 204
Mercuric fulminate, 280

Nickel carbonyl, 202
Nitric acid, 151
— oxide, 146
Nitrogen, 117
— combination of, with magnesium, 124
— — — — oxygen, 122
— combustion of lithium in, 126
— iodide, 142
— — action of light upon, 143
— oxides of, 143
— pentoxide, 151
— peroxide, 150
— trioxide, 149
Nitrous chloride, 141
— oxide, 143
— — liquid, 144
— — solubility of, in water, 333

OER

Oersted condenser, 301
Oxides of nitrogen, 143
— and acids of chlorine, 100
— — oxyacids of iodine, 113
Oxygen, 20
— liquefaction of, 307
— solubility in water, 329
Ozone, 29
— absorption of, by turpentine, 35
— test papers, 30, 38

Perchloric acid, 104
Periodic classification of the elements, 343
Phosphonium bromide, 288
— chloride, 290
— iodide, 251, 253
Phosphoretted hydrogen, 249
— — action of iodine upon, 252
Phosphorous oxide, 255
Phosphorus, 242
— glow of, 245
— pentoxide, 254
— red modification of, 247
— spontaneous inflammability of, 242
— suspended solidification of, 243
Platinum tube, 5
Pneumatic troughs, 6
Pyrophorus, 176, 181

Salts, solubility of, in water, 64, 330
Sea-water, composition of, 334
Siemens ozone tube, modifications of, 31, 32
Silicic acid, 239
Silicon, 238
— fluoride, 116
— hydride, 238
Silver hydrozoate, 141
Silvering glass, solutions for, 325
Singing flame, 15
Soap destroyed by various waters, 333
Soap-bubbles, solution for, 10
Specific heat of elements, 336
Steel mill, 173

XIN

Sulphur, 258
— combustion of oxygen in, 171
— dioxide, 267
— — combustion of iron in, 270
— — composition of, 269
— — decomposition of, by light, 269
— — liquefaction of, 268, 296
— — solubility of, in water, 269, 333
— plastic, 259
— prismatic, 259
— trioxide, 271
Sulphuretted hydrogen, 262
— — absorption of, by charcoal, 182
— — action of, upon metallic salts, 266
— — synthetical formation of, 262
Sulphuric acid, 273
— — combination with water, 274
— — action on sugar, 275
— — — — organic matter, 275
— — table of strength of, 340

Tension of aqueous vapour, 331
Turpentine, absorption of ozone by, 35

Water, 38
— colour of, 48
— electrolysis of, 43
— freezing of, 54, 58
— maximum density of, 51
— of crystallisation, 67
— solvent action upon gases, 71
— — — — liquids, 68
— — — — solids, 64
— specific heat of, 53
— spheroidal state of, 50
— super-cooling of, 55
— synthetical formation of, 39
— volume composition of, 40
Waters, analysis of natural, 335
— classification of, in order of softness, 333

Zinc-copper couple, 2, 205
Zinc ethyl, 175

Printed in England at THE BALLANTYNE PRESS
SPOTTISWOODE, BALLANTYNE & CO. LTD.
Colchester, London & Eton